Aseptic Processing and Packaging of Particulate Foods

Aseptic Processing and Packaging of Particulate Foods

Edited by

EDWARD M. A. WILLHOFT
Principal
Epsom Technical Group Services
Epsom
Surrey

SPRINGER-SCIENCE+BUSINESS MEDIA, B.V

First edition 1993

Originally published by Chapman & Hall in 1993
Softcover reprint of the hardcover 1st edition 1993
Typeset in 10/12 pt Times New Roman by DSC Corporation, Cornwall, England

ISBN 978-1-4613-6361-3   ISBN 978-1-4615-3112-8 (eBook)
DOI 10.1007/978-1-4615-3112-8

A catalogue record for this book is available from the British Library

Preface

Publications in food technology proliferate; however, noticeable by its absence of coverage is the subject of processing and packaging of particulates in foods. Recent years have seen significant advances which will almost certainly result in substitution of existing and conventional retorting. In addition, when combined with high temperature/short time (HTST) processing, we can expect substantial further growth, reflecting quality and convenience advantages over products processed from yesterday's technologies. The anticipated growth in particulates is driven by both materials and packaging advances and only requires modest marketing of the organoleptic advantages to establish their place on menu options.

The directions taken in packaging developments, especially those interfacing with the latest and established methods of processing, are increasingly influenced by the need to design packaging on a cradle-to-grave basis. Time was when multi-laminated films on board satisfied the total needs of consumers of aseptic products. The problems of recycling combustible, i.e. energy generating materials laminated with aluminium foil, are becoming sensitive issues in a world preoccupied with recycling, and are creating openings for alternative and environmentally friendly material combinations.

This book brings together advanced technologies in the field, to provide information for professionals with interests in aseptic processing on how to go about selecting a system appropriate to their commercial needs and constraints. It covers markets for products, processing and materials in the USA, Europe and Japan. It avoids or minimizes duplication with works in similar fields and emphasizes the technologies behind processing plant, especially those for handling particulates. Integration of processing with packaging is discussed, including the latest proven technologies based on microwave cooking, resistive heating and neutral aseptic handling of packaging. The book also includes an appreciation of packaging requirements, microbiological aspects applied to aseptic processing, and critical control point hazard analysis (HACCP). The efficacy of HTST is highlighted in terms of heightened sensory quality and improved sterility.

The Dole process is included because it has stood the test of time and was an early adjunct to HTST. It is capable of being developed for particulates and, perhaps as importantly, with the emphasis on recycling tin-plate or its equivalent, is readily recyclable without unduly disturbing the environment.

This book is recommended reading for food and packaging technologists, production and process engineers, quality control management, material convertors, microbiologists and, last but not least, market research and marketing personnel of food manufacturing companies and trading operations.

Finally, I should like to thank each one of the authors for the time, patience and commitment that they put into presenting their contributions, and especially Margaret for her unstinting support throughout.

E. M. A. W.

Contents

Contributors

Professor Dr N. Buchner	University of Stuttgart-Hohenheim, Postfach 70 0562, Schloss 1, 7000 Stuttgart 70 (Hohenheim), Germany
Mr R. E. Coles	Foodpack Information Centre, 16 Kenilworth Road, Cubbington, Leamington Spa, Warwick, UK
Mr D. A. Joyce	David A. Joyce Associates, Manley Common, Cheshire WA6 9ES, UK
Mr C. J. Robinson	Rémy Gimpel Ltd, 103–109 Lavender Hill, London SW11 5QI, UK
Dr P. J. Skudder	APV Baker Ltd, Automated Processes Division, PO Box 4, Gatwick Road, Crawley, Sussex RH4 2QB, UK
Mr B. I. Turtle	45 Priory Orchard, Wantage, OX12 9EL, UK
Mr S. Wakabayashi	Shikoku Kakoki Co Ltd, Toyo Bldg, No 20, 12-banchi, Jingu-mae 6-chome, Shibuya-ku, Tokyo 150, Japan
Dr E. M. Willhoft	Epsom Technical Group Services, 41 Higher Green, Epsom, Surrey KT17 3BB, UK
Mr F. S. White	Graham Engineering Corp, 1420 Sixth Avenue, PO Box 1140, York, Pennsylvania 17405, USA

1 Aseptic processing and packaging of food particulates

N. BUCHNER

1.1 Introduction and basic principles

Aseptic processing and filling consists of presterilization of the product before filling the sterile product into a sterile—usually also presterilized—package in a sterile atmosphere with sterile mechanical means and subsequent closing of the package in a sterile manner. After the packaging process no further sterilizing influence is exerted from the packaged product, the packaging material, the interior atmosphere, or from outside the package. A so-called 'commercial sterility' has to be achieved with usually a maximum of one non-sterile package out of 10 000, allowing for conditions of storage and distribution.

Aseptic processing and packaging has some advantages over competing methods of food preservation, but it has also some disadvantages. Aseptic processing and filling is associated with high capital expenditure on plant used for presterilizing and filling of products and especially for automatic control of all conditions safeguarding product sterility; all steps controlling sterility should be controlled through automation. Limitations for aseptic processing and packaging still exist in the areas of presterilization and filling of difficult products; for example, liquids with large particles. Developments, however, are proceeding and these limitations are gradually disappearing.

1.1.1 Advantages over in-container sterilized foods

Quality of aseptically treated foods may be higher due to UHT (ultra high temperature/short time) sterilization. This is especially valid with large packs, because heat penetration in retorting is time-consuming and consequently quality suffers from overheating. UHT treatment is usually performed with processes allowing fast heating and cooling; for example with thin layers of product in heat exchangers. High UHT temperatures make use of the fact that the sensitivity of microorganisms to high temperature is much higher than that of the food constituents. While the Q_{10}-value (factor for speeding up reactions when raising temperature by 10° C) for chemical reactions of food constituents is usually between 2 and 4, the equivalent value for killing bacteria is 10. If the temperature from the usual retorting value of 121°C is raised by 10°C, the time for the same killing rate for bacteria will be 10%, and with a 20°C rise only 1%. If for chemical reactions a Q_{10} of 3 is assumed, they will be speeded up by only a factor of $3 \times 3 = 9$ when raising the temperature by 20°C. As a result, a temperature rise from 121°C to 141°C for sterilizing foods will give theoretically only about 9% of the damage to quality, compared with retorting at 121°C.

Usually less energy is needed for UHT sterilization because at least part of the process heat may be recovered with heat exchangers. Recovery rates may be higher than 90%. Also, the aseptic process is a modern continuous flow process, needing fewer operators and space than retorting. At the discharge of the aseptic filler the packages are ready for storage. Packs do not require to be transferred into retorts where, depending on the nature of the retorts, certain expenditures are necessary for handling the packs. Packaging material and packs are sterilized under milder conditions than those existing in retorts and therefore a wider choice of packaging materials and packs is possible: packs may be cheaper, lighter, less space-consuming and offer more convenience.

Compared with retorted trays and cups of polypropylene/ethylene vinyl alcohol/polypropylene (PP/EVOH/PP) used for soups and other meals, aseptically filled packs from these materials do not suffer during subsequent storage from water vapour absorption by EVOH during retorting and raised oxygen transfer from the atmosphere. Even after drying the permeability of retorted EVOH may remain higher due to 'retort-shock'.

1.1.2 Advantages over pasteurized chilled foods

The nature of the products and processes determines whether the quality of aseptically filled products may be lower than that of pasteurized chilled foods. Aseptically filled products, however, have some advantages, if filled and sealed effectively. They are microbiologically safer, because they need no refrigeration. No control is necessary for possible interruptions of the chilling chain, which, in the case of the short shelf-life of pasteurized products, creates a cost penalty.

As aseptic products do not need refrigeration, storage and distribution are cheaper and selling paths may be longer and permit more variation. Production batches may also be larger: so production becomes more economical. If a modified atmosphere is used for chilled products, often larger packs have to be used to contain sufficient CO_2 as a control medium for microorganisms.

1.1.3 Advantages over 'cleanfill'

Cleanfill is a process similar to aseptic filling. Total commercial sterility, however, is not the target of cleanfill, since a very low level of microorganisms remains after filling and sealing, allowing a shelf-life of 4–6 weeks when products are chilled. While the costs for cleanfill are somewhat lower than those for aseptic filling, and the quality image may be somewhat higher due to refrigeration, disadvantages are similar to those for pasteurized chilled products, with the exception that the microbiological risk is lower provided that only high-acid foods are cleanfilled.

1.1.4 Advantages over frozen products

Frozen foods have a long shelf-life, high quality and a low microbiological risk (if the cooling chain is not interrupted) similar to aseptically treated products. Freezing is suitable for a wide variety of products that cannot be aseptically filled, because of small production batches and difficulties of handling. Where, however, aseptic products are highly competitive, frozen foods demonstrate disadvantages: for example they require cooking whereas aseptic products are generally ready to eat or drink. Where products are heated for consumption little time is needed for aseptic products in contrast to frozen foods. In microwave ovens ice absorbs only about 0.1% of the energy absorbed by water or water containing unfrozen food (Schubert, 1990). This results in an extended heating time for frozen foods with very uneven heating and overheating of melted zones.

Aseptic products may also be distributed to countries with a poor cold storage infrastructure. Maintaining quality of frozen foods during storage and distribution has an extremely high energy demand. More than 60% of the energy consumption in the food system, starting with agriculture or fishery and ending with the packaged product in the hands of the consumer, is needed for cooling during storage and distribution (Khaladadij-Nia and List, 1981).

1.1.5 Advantages over dried products

Advantages of dried products are long shelf-life and absence of microbiological risk, weight reduction by drying and low-cost packs. Quality preservation by drying, however, is limited to foods suitable for drying. Quality and consistency of dried products may suffer by the drying process. In contrast to aseptic products, dried products need to be reconstituted for consumption and so their convenience is lower.

1.2 Developments in aseptic processing and filling

Aseptic methods have a history of nearly 80 years. In 1914, sterile filters for filtering wine were fully developed. In 1917, Dunkley in the USA received a patent for sterilizing cans and lids with saturated steam and subsequent aseptic filling. In Denmark and the USA processes and machines for aseptic filling were developed around 1920, but all these processes disappeared either from lack of reliability or flexibility. It could, however, be demonstrated that high-temperature/short-time treated milk had a superior quality compared with the retorted product (Emch, 1978).

In the early 1940s, the so-called Dole–Martin process was developed by Dr Martin, USA; indeed, machines for this process are still manufactured and supplied today. In 1962, the development of Alpura AG (Löliger) in Switzerland for the aseptic filling of milk in a modified Tetra Pak machine to produce the tetrahedron packs from carton/polyethylene was completed, and the Swedish company, Tetra Pak, started marketing the process on a licence basis. This development was the beginning of the modern basis of aseptic filling, because it gave the impetus for expanding the method to different package types, such as bags, cups, different carton styles and, later, also bottles. This method usually applied sterilization with hydrogen peroxide as in the Alpura system but with many modifications. In 1969 the Tetra Brick aseptic machine was introduced and, in the late 1970s, the aseptic Combibloc system using carton blanks instead of roll material was introduced.

An important date for the expansion of aseptic methods was January 1981. After animal tests with high levels of hydrogen peroxide this sterilizing medium had been attacked in Japan and, in January 1981, the US Food and Drug Administration (FDA) allowed sterilization of polyethylene surfaces (and later different plastics) as a safe method—after a thorough study of the Japanese findings. This gave a great impetus to aseptic methods, not only in the USA but throughout the world. Packaging pastes with particles started gradually, when small particles of fruits or rice or tapioca were added to yogurt or pudding. In the mid-1970s, much development work was performed for processing and filling fruit purées and particulates in drums (Loire, 1976) and later in bag-in-box packs. The scientific basis of those developments is described elsewhere (Kessler and Fiedler, 1985). In 1984–85 the first Combibloc packs containing aseptically processed soups entered the British and, later, other markets. New items such as tomato products followed (Mankowski, 1987).

Much development has been achieved in the last few years in filling soups and stews aseptically, for instance, into cups and trays suitable for microwave heating, and this is on the verge of commercial application.

1.3 Aseptic processing—methodology

Aseptic processing comprises the following:

1. Sterilization of the products before filling.
2. Sterilization of packaging materials or containers and closures before filling.
3. Sterilization of aseptic installations before operation (UHT unit, lines for products, sterile air and gases, filler and relevant machine zones).
4. Maintaining sterility in this total system during operation; sterilization of all media entering the system, like air, gases, sterile water.
5. Production of hermetic packages.

1.3.1 Sterilization of products

Presterilization of products consists usually of heating the product to the desired UHT temperature, maintaining this temperature for a given period in order to achieve the desired degree of sterility, with subsequent cooling, usually to ambient temperature and sometimes to an elevated temperature to achieve the right viscosity for filling. Heating and cooling should be performed as rapidly as possible to achieve the best quality dependent upon the nature of the product. A fast heat exchange rate is desired for cost reasons.

1.3.1.1 Methods for liquid products These are as follows.

Sterile filtration. As mentioned already sterile filters were developed very early. They are used for clear liquids such as wine and fruit beverages (Haubs, 1988) and also for pharmaceutical preparations, where heat may not be applied (Wallhauser, 1988), and for sterile water if needed in sterile installations. These sterile filters have to be protected from even small particles by prefilters. In the last 15 years great progress has been made with filter materials, their construction and in-line testing for integrity, with the result that product sterility from filtration is now comparable with thermal sterilization methods.

Plate-type heat exchanger. Plate heat exchangers have been used for sterilizing with subsequent hot filling of fruit juices for a long time. When applied to aseptic filling they serve both for heating (to higher temperatures) and cooling. The advantages of plate heat exchangers are a high heat-exchange rate due to the large exposed surfaces and the generation of turbulent flow at moderate costs; disadvantages are intolerance of gaskets to pressure and of the whole system to even small particles such as fibres. Therefore standard plate-type heat exchangers with mechanical contacts with the plates may be recommended for clear liquids only.

Steam injection in liquid. The most rapid method of heating liquid products is by injection of steam. Within fractions of a second the desired sterilization temperature may be attained. Combined with the most rapid methods of cooling such as injection of the hot product into a vacuum chamber and evaporation of an equivalent amount of water, a very high quality level may be achieved. The method is suitable, however, for particle-free liquids only. When sterilizing particulate mixtures, it may be used for sterilizing the liquid separately as the particles have to be sterilized using a different method. A patent was applied for this process by Alpura AG, Switzerland, in 1949 (Emch, 1991). The method is combined usually with heating and cooling in heat exchangers to the lower temperature range, for instance below 80°C. A disadvantage of the system is the reduced heat recovery rate of about 50%.

Liquid infusion into steam. A similar method is infusion of a thin film of liquid into a steam atmosphere, which also gives rapid heating. One type infuses a thin flat film; with a newer method a thin circular film may be infused, avoiding thicker rims at the ends of the flat film requiring longer heating times. Cooling is also achieved by infusion of the liquid into a vacuum chamber and evaporation of water. Advantages in quality, range of application, preheating and heat recovery rates are similar to the method of steam injection.

Friction heating. A method of heating liquids by friction between rotating and static disks has been developed, which combines heating with some homogenization. The method, suitable for liquids only, seems not to be applied commercially.

1.3.1.2 Methods suitable for liquids with particulates. These are as follows.

Plate-type heat exchangers with free-flowing plates. These heat exchangers with mechanically-stable, free-flowing and especially contoured plates without mechanical interconnections (except at the rims), have good heat-exchange rates due to large exposed surfaces and turbulent flow characteristics and, for this reason, they are applied to UHT treatment of liquids with smaller particulates up to 5–7 mm, or about half of the distance between the plates (Buchwald, 1987).

Tube-type heat exchangers. This type of heat exchanger has the widest field of application for UHT sterilization of liquids and pastes, also containing small particulates. The design of the apparatus is limited to a maximum size for particulates to be handled. There are cylindrical tube-type heat exchangers and others with straight tubes. In the bows at the ends, turbulence is created, which mixes product and gives short heating and cooling times. Sizes of particulates are equivalent to cubes of 7–8 mm maximum, depending on the design. If one dimension of a particulate is smaller, others may be larger.

Scrape-type heat exchangers. Scrape-type heat exchangers are similar to straight-tube heat exchangers, but they have a central rotating shaft carrying scraping devices for the heated surfaces to prevent burning and fouling of foods at the surfaces as well as providing a mixing action. These heat exchangers are suitable for liquids with high viscosity and also for liquids and pastes with particulates up to about 12 mm cube; although tests with specially designed and optimized versions have shown that particulates up to 37 mm could be processed. Scrape type heat exchangers have been used for a long time for heating and cooling, and also for UHT treatment prior to conventional filling, followed by retorting. Much basic work has been done to provide a better understanding of flow and heat-exchange characteristics (Ohlsson, 1988; Wernimont, 1988).

Rotaholder. A system, being basically a tube sterilizer in which liquid and particles are flowing jointly, was developed with the addition of 'Rotaholders' in which the particulates are held back for a certain period by rotating holders with fork-devices through which the liquid may flow. This compensates for the longer time needed by particulates for heat penetration during heating and cooling. The heat load for particulates and liquid may thus be adjusted to achieve a good overall quality. Size limitation is 15-18 mm cubes with a maximum of 15% particulate content in the liquid. The viscosity of the liquid is limited by the necessity of a free flow around the particulates in the holding device. This method is at an advanced development stage (Hermans, 1988).

Sterilization of particulates in a fluidized steam bed. In Australia a system ('Steriglen') was developed in which particulates, in the test meat and fish particulates were used, were separated from liquid, which was individually sterilized, and the particulates were introduced into a pressurized chamber through a pressure lock. In the chamber the particulates were heated in a fluidized bed by steam and then cooled with sterile nitrogen flowing through the layers of particulates. The cooled product was transported out of the system through a second pressure lock. The development was transferred to the USA and seems not to be commercially exploited.

Double-cone sterilizer ('Jupiter'). In England a semi-continuous system was developed in which most of the liquid is sterilized continuously while the particulates, with a small amount of liquid, are sterilized discontinuously in a slowly rotating double-cone pressure vessel with an interior steam atmosphere. The jacketed walls are either heated with steam or cooled with water. During heating preheated liquid may be injected to speed up heating. After cooling the particulates and liquid are usually mixed again in a stirred mixing tank. Separate filling is possible (Hastings and Murray, 1987). The advantage of this system is that particulates and most of the liquid are sterilized individually according to their characteristics; disadvantages are the high costs involved for separating the components, the requirement of two processes (continuous and batch), and remixing, also possible quality changes by the mixing and agitation. A totally continuous process would be more desirable.

Sterilizing by microwaves. Microwave heating of food may give rather rapid heating, because heat is created in the food directly. Heating, however, is not totally uniform because of differences in dielectric constants and specific heat values of the different food constituents. Some time for equalizing temperatures by heat conduction has to be allowed. Microwave energy may enter into the flowing food through tubes from materials transparent to microwave energy, for instance from Teflon, polypropylene or ceramics. Microwave generators with high efficiency (83%) and high power output now exist (Meredith, 1988).

In Japan, a continuously operating microwave sterilizer has been developed and tested (Iwamato, 1985). It is not known whether a commercial installation is operating.

One advantage of microwave sterilization is fast and uniform heating of particulate mixtures. As heat is created directly in the food particle, sizes are limited only by valves, pumping devices and fillers. Mechanical damage to particulates is low as food is flowing as a plug; the cost of microwave generators, however, is high. Cooling is conventional, in other words time-consuming with possible negative consequences for food quality. Heat recovery exists for the conventional first heating section; electrically applied energy may be recovered only as thermal energy.

1.3.1.3 Electrical resistance heating for sterilizing. With this method the food provides the electrical resistance and is directly heated resistively. Investigations into this method were first performed in 1910. Development work had to be abandoned because of corrosion problems at the electrodes. In the recent developments these problems—and that of electrolysis of the product—could be eliminated by proper selection of electrode materials and frequencies.

Recent developments were carried out in Finland (Anon., 1981) for sterilizing milk and in England for particulate foods (Skudder, 1988). Research was also carried out in Germany (Reitler, 1990). Limitations similar to those for microwave sterilization apply to this method. Muscle, for instance, has a similar electrical resistance to food fluids; fat, however, may not be heated directly. For this reason, sufficient time for heat conduction is necessary to achieve sufficiently uniform temperatures. As the electrical conductivity of some food constituents is low below 80°C, it is best to heat foods to this temperature conventionally and then the remaining increase to sterilizing temperatures, by electrical heating.

Advantages and disadvantages are similar to those for microwave sterilization, except for one important difference: magnetrons for microwave heating are costly, while installations for electric resistive heating are, in comparison, inexpensive. The efficiency for direct electric heating is practically 100%.

High-voltage discharge (Elsteril). It is already known since the end of the nineteenth century that bacteria may be killed by high-voltage discharge from loaded capacitors: their membrane is destroyed. This method was recently investigated in Germany (Sitzmann and Münch, 1990). With a 20 kV and pulsed discharge, at certain frequencies and for certain microorganisms a kill rate of 5 decimal values was achieved. For foods, however, and *Clostridium botulinum* a kill rate of 12 decimal values is necessary. This is a critical demand from the process, because about 28 kV/cm would be needed and this value is higher than the breakdown voltage for wet food. Also this process cannot be used for large particulates, because to achieve the necessary electric field strength, the electrodes would be too close. Therefore this interesting process, which implies virtually no heat applied to the food, seems to be adequate only for pasteurization and reduction of microbiological loads of foods in a continuous process.

Ultra-high-pressure sterilization. It has been known for a long time that microorganisms may be killed by ultra-high pressures of thousands of bar. Some years ago at Kyoto University in Japan, investigations in this field were started. It was found that a certain rise of temperature, for instance to 45°C, could bring down the needed pressure to 3000–4000 bar. *Bacillus cereus,* however, needed temperatures of about 65°C. The process itself gives only a very moderate temperature rise and foods retain their natural taste. Protein, however, is coagulated and so foods may change their visual appearance. This could be the basis of creating a new class of high-protein foods (Hayashi, 1989). Commercial installations have been built for high-acid products only, such as jam and fruit juice, which are less critical to sterilize.

1.3.2 Sterilization of packaging materials and packages

Microbiological count of films and packing. The level of microbial contamination of packaging materials stored in rolls is low, because the individual layers protect one another.

The initial contamination of packaging material with extruded plastic upper layers or with heat-seal lacquer, which is then heated for drying, is low. Despite the fact that it was shown that microorganisms may survive extrusion conditions (Voss and Moltzen, 1973) use of plastic granules with low microbiological count and adequate extrusion conditions gives a sufficiently sterile extruded plastic.

The microorganisms on packaging materials derive from sedimentation from the atmosphere. The count is between 1 and 10 microorganisms/dm^2 and with good manufacturing conditions, closer to 0.1–0.2 mo/dm^2 (Swartling and Lindgreen, 1962; Lubieniecki, 1973; Placek and Witter, 1972; Witter, 1979). Similar values are found on dried heat-seal lacquer (Sturm and Gilland, 1974). Here the individual groups of microorganisms have also been monitored: bacilli accounted for only 3%. Microbiological load and type of microorganisms vary for different seasons. In summer the percentage of moulds and yeasts is higher. With preformed plastic cups, values were found to be about 10-fold (Vossen and Moltzen, 1973). Since difficulties with these preformed cups may arise from the high load of microorganism and also from dust, precautions need to be taken to keep these contaminants at a low level (manufacturing and printing in clean areas, no contact with hands, protecting by polyethylene bags). Although glass bottles and jars are sterile when manufactured, during the last step of cooling, however, infection may occur by airborne microorganisms. The infection rate is dependent on the size of the opening of the bottle, level of contamination of the air and the use of mechanical precautions. With bottles having an opening of 20 mm, values of 2–4 were found as an average immediately after manufacture. After storage for 2 months these values were reduced to about 20% of the initial count (Streubl, 1988). Arrangement on pallets has to be done under good conditions. Openings have to be protected by plastic layers during shipping and storage (Buchner, 1989). Bottles from plastics have a very low count of microorganisms, if properly manufactured, handled and protected, for instance in polyethylene bags.

Requirements for low and high acid foods. These are as follows:

Low-acid foods (pH > 4.5). All types of microorganisms may grow in low-acid foods. If a maximum of 1 non-sterile package out of 10 000 has to be achieved (for baby food the demand is still higher) and packaging material alone is regarded as a source of infection, in 10 000 packages 1 microorganism is allowed to survive. The kill rate necessary for this target is dependent on the initial count of microorganisms and the area needed for a pack. If this area is 1 dm^2 and the load 1 mo/dm^2, 10 000 packages carry 10 000 mo and the kill rate has to be 4 decimal values in order to reduce to 1 mo. For small minicups with 0.1 dm^2 surface, 3 decimal values are sufficient for the same result; for packages with 10 dm^2 surface, for example milk packages, a 5 decimal value would be necessary. The US FDA usually demands 5 decimal values for packages with about 1–10 dm^2 surface. For *Clostridium botulinum* a 10 decimal value is demanded by the FDA for packaging materials. This would mean a 4 decimal requirement when sterilizing with hydrogen peroxide, because *C. botulinum* has a 2.5-fold kill rate in the logarithmic system compared with *B. subtilis globigii*, which is usually used to test hydrogen peroxide sterilizing systems as it is the most resistant strain to this medium (Denny, Brown and Yao, 1974; Cerf and Hermier, 1972). Determining the required kill rates with the most resistant microorganism gives additional safety, because all other microorganisms are killed faster. This also compensates for small contributions to non-sterility arising for instance from the food or air.

High-acid foods (pH < 4.5). In high-acid foods only part of the microorganisms found at the surface of packaging materials is able to grow, usually less than 10%. Therefore the calculation could be performed with a smaller number of microorganisms than found on the material. Also the sensitivity of these microorganisms to sterilizing influences is much higher than that of *B. subtilis*. For sterilization with dry and moist heat for high-acid filling it was found that *Aspergillus niger* is a suitable test organism because it showed the highest resistance (Thum, 1982).

1.4 Maintaining sterility in aseptic machines

Aseptic machines have to be sterilized prior to operation and kept in sterile condition during operation. There are various existing options: protection of relevant aseptic zones by steam or by sterile air or gases (Buchner, 1991).

Protecting sterile zones by steam is performed within the whole aseptic machine in the Dole machine, whereas protection of the filling zone only is a usual method with bag-in-box systems.

For protection of sterile zones by sterile air/gases three different methods exist:

1. Protection by sterile air flushing of the opening of the package during filling and sealing; this method is used in the Bottlepak system. An extruded tube of plastic is blown to the bottle shape with sterile air and, during filling and sealing, the opening of the pack is protected against infection by flushing this zone with sterile air.
2. Protection of larger zones in packaging machines starting with sterilization of the pack from filling up to sealing by a vertical laminar flow of sterile air. This method is used for bottles, cartons, cups and heat-sealed composite cans. It is simple and has some limitations: machine elements above the openings of the packages, which have to be sterile, may create adverse turbulences, especially if moved, which can give a backflow of non-sterile air and infection of the packs. Turbulences may also occur at the rim of the

packs: a design that gives laminar flow here and/or outside sterilization of relevant package areas is necessary.

3. Application of totally closed machine cabinets with turbulent ventilation and slight overpressure. All surfaces, also those of packs, must be sterile because the microorganisms may be carried around. One advantage of this system is the very high microbiological safety level. Another is the great variety of possibilities in design and movement of machine elements. There are also advantages for presterilization, as mentioned later.

1.4.1 Sterilization of aseptic installations

Aseptic installations have to be sterilized prior to production, at least those parts from which infection could occur in the packaged food. The US FDA demands a 4 decimal kill rate for *B. subtilis globigii* in low-acid machines when sterilizing with hydrogen peroxide. There are three main fields of concern: packaging machines, filling lines and fillers, and air/gas systems. Filling lines and fillers have to be cleaned automatically prior to sterilization in the clean-in-place (CIP) process. The process of presterilization of all three items should be automatically controlled.

1.4.1.1 Sterilization of packaging machines. In machines in which steam is used as protection against infection, steam also serves to sterilize relevant zones of the machines.

In open, sterile air systems such as local flushing or laminar flow systems, it is difficult to prevent sterilizing vapours from escaping into the working room. Specialised devices then become necessary. In closed systems few precautions are necessary to prevent these vapours from escaping. Vapours may be centrally discharged and, in the case of hydrogen peroxide vapours, also decomposed catalytically to oxygen and water vapour.

The methods of sterilization are as follows:

1. Hydrogen peroxide is used in many cases as spray or vapour, which is condensed at the surfaces of the machine with subsequent drying.
2. Various disinfectant solutions for sterilization of packaging materials are used also for presterilization of machines.
3. Hot air is used in various systems such as Erca and Tetra Pak (330–350°C, for 30 min), requiring high temperatures and long sterilization periods.
4. Water vapour for presterilization of machine zones is used in the Dole system (superheated atmospheric steam for low-acid packaging) and in two Bosch systems (atmospheric wet steam) for high-acid beverages, the Hypa-S system and the aseptic bottle system.

1.4.1.2 Sterilization of filling lines and fillers. For this purpose usually pressurized steam or pressurized superheated water is used, typically at temperatures of

130–135°C corresponding to an overpressure of about 2 bar, whereby filling lines and filler usually are sterilized together with the UHT unit. In order to achieve reliable sterility cold spots, for instance from included air, have to be avoided. Time for sterilization is determined by the temperature of the coldest spot that has to be controlled automatically. In a French system, filling lines and filler are sterilized without overpressure by polyalcohol mixtures at 170°C, which have to be cleaned off thoroughly after sterilization by rinsing with sterile water.

1.4.1.3 Sterilization of air filters. Most systems create sterile air by filtering; a few systems also by incineration (heating the air to 300–350°C and cooling). Filters should be sterilized prior to operation and at regular intervals if there is danger that microorganisms could grow in the filter material, which is dependent on relative humidity. Cartridge filters of different design and materials may be sterilized by pressurized steam; large surface HEPA filters for filtration in laminar flow systems, however, are unsuitable for this process. In some cases they are sterilized by proprietary processes. In most cases, however, they are not sterilized but only blown clean which creates a possible problem when starting, or after repair operations or replacement of filters. In order to avoid contamination, filters have to be vented continuously.

1.5 Filling

Fillers have to handle the product gently. They must be suitable for CIP and SIP (sterilize-in-place) as mentioned. With some systems, such as piston fillers, elements like pistons and valves are redrawn to allow access for cleaning and sterilizing media to all surfaces.

1.5.1 Filling principles for liquids

For products that are difficult to handle, such as particulates in a low-viscosity liquid with the inherent tendency to demix, it may be reasonable to fill particulates with little liquid in a separate filler. A special two-step filler deposits particulates with a little liquid in the first stage: liquid fill in the second stage then cleans the nozzle.

1.5.1.1 Package controlling the filling volume. In the usual Tetra Pak, and similar systems that manufacture a tube of packaging material from which packages are separated by sealing and cutting, a column of liquid is kept constant in the tube and final sealing of the packs is through the liquid. Thus the pack itself controls the filling volume. Deviations are within legal limits. This method is suitable for liquids without particulates only.

1.5.1.2 Diaphragm filler. For small fill volumes, such as cream cups, diaphragm fillers with a flat membrane are used. For larger filling volumes tube-type diaphragms are used; these are compressed and expanded by the action of a piston pump with glycerine surrounding the tubes. A special piston filler uses a rolling diaphragm as a barrier between sterile liquid and non-sterile outside air.

Diaphragm fillers are unsuitable for liquids with large particulates but small and soft (non-abrasive) particulates in the liquid may be handled, depending on the design of the valves.

1.5.1.3 Time-lapse filler. For special purposes, such as filling small amounts of liquid, fillers are used in which volume is controlled by adjusting the time of filling. This simple device is sensitive to certain external influences, such as variations of pressure or height of liquid in the vessel, or viscosity (depending on temperature), and has, for this reason, little application. As flow characteristics would be influenced by particulates this principle is unsuitable for such mixtures.

1.5.1.4 Induction flowmeter. If electrically conducting liquids pass through a magnetic field they induce an electrical voltage. This effect is used for measuring the flow velocity of liquids. The flowmeters actuate valves when a preset volume has passed. Induction flowmeters are used for aseptic filling of liquids such as milk and fruit juices. Larger particulates in the liquid would give variations of conductivity and volume. The advantage of this system is smooth flow through the measuring tube.

1.5.1.5 Piston fillers. Piston fillers are also used for filling liquids and pastes such as cream, yogurt and pudding. Particulate filling is dependent to a large extent on the design of the valves. Non-sterility from the mechanical drive is prevented by various methods: with the Bosch filler internal sterile air and the external non-sterile air are separated by rolling diaphragms.

1.5.2 Filling principles for particulate mixtures

1.5.2.1 Piston filler. Piston fillers are presently used to fill particles up to a size of 25 mm. Smooth design of the path of the food to and from the pistons has to guarantee uniform flow without blocking and demixing. For filling nozzles and valves, many proprietary designs are used in order to achieve filling without splashing and dripping. Rotating and reciprocating valves are used. Valves and nozzles are adjusted to the demands of the products. A special piston filler (Raque) applies a plunger behind the outlet valve to push down the product into the package.

1.5.2.2 Rotating sine pump. Rotating sine pumps give a uniform flow of product. At both sides of the rotor sine-shaped pockets exist, limited by a reciprocating slider. During rotation one product-filled pocket becomes smaller and the opposite one grows in size. Particle size is decided by the size of the pockets and, thus, of the rotor. Large pumps have large pockets, but they are suitable for larger amounts of product, too. For small amounts the variation in preselected rotations would give variations in volume that are too large. A typical application is in the Bosch vertical aseptic form/fill/seal system: pouches with volumes of 1–5 l can be filled with food with particles up to 12 mm.

1.5.2.3 Pumps with two rotating lobes. Suitable pumps with two rotating lobes —for instance from Waukesha—may be used in a similar way to sine pumps. Damage to product depends on size and characteristics of product and the design of pump and valves.

1.5.2.4 Mass flowmeters. Mass flowmeters, measuring the flow by the Coriolis effect, are used in bag-in-box systems to determine the filled volume. This is preset and the flowmeter actuates the valves. Mass throughput is measured independently of the percentage of liquid and particulates. The design of the flowmeter, the valves and the pumping unit controls the maximum size of particulates that may be handled without restriction or deterioration of the product.

1.6 Closing packs

Faults in closing packs are the most frequent cause of non-sterility. For this reason, perfect closing and integrity control have the highest priority. The closing operation often is linked with headspace gas, or a steam-flushing action. Different methods have been developed for this purpose: bulk gas flushing of webs with formed and filled cups in thermoform/fill/seal machines or individual flushing of sealed composite cans (Hypa-S) or bottles or gas- or steam-flushing of cartons or glass bottles. In vertical form/fill/seal systems, as with the Bosch vertical pouch system or the Tetra Brick headspace system, the whole tube above the product may be filled with sterile neutral gas, which gives very low oxygen concentrations in the headspace gas.

1.6.1 Heat-sealing

Heat-sealing with heated and temperature-regulated sealing bars and plates is the most common method of closing modern aseptic packs. A precondition for a good result is that sealing areas are free of large particles of food stemming from filling or splashing during transport. The sealing tools are quite different according to their tasks. Temperature and pressure distribution has to be uniform, tools have to be non-corrosive (also free from chemicals for pre-sterilization of machines) and tool and packaging materials have to be selected in a way to avoid sticking to the tools during sealing. There are sealing tools for individual sealing of bags, cups and plastic bottles and sealing plates that seal up to 84 cream cups at one stroke, where the counterpart tool adjusts itself by spring pressure.

1.6.2 Types of packs

1.6.2.1 Consumer packaging. A great variety of packages may be aseptically filled now as listed:

1. Carton boxes. Some of the existing aseptic carton boxes may now be filled with particulates, also aseptically.
2. Bags and pouches. Pillow pouches are usually used for packaging milk; one three-sided sealed pouch, however, is suitable also for aseptic packaging of particulates up to particle sizes of 12 mm and bag sizes from 1–5 l. For standing pouches a Japanese machine (Fujimori Kogyo) uses closed pouches from a reel with sterile interior surfaces, the exterior of which is sterilized in a hydrogen peroxide bath when the web with pouches enters the aseptic cabinet. The bags are then cut from the web, filled and sealed.
3. Cups and trays. These are either used premade or formed, filled and sealed in thermoform/fill/seal machines. Both types of machines exist for filling particulates and also in packs suitable for microwave heating. Usually polypropylene-based multilayer materials with EVOH barrier are applied for this purpose.
4. Bottles and jars. Glass bottles may be aseptically filled with food containing small particles, for instance for baby food. Jars may be filled with larger particles—12 mm cube size or larger—if one dimension is smaller. In a recent development returnable bottles are filled aseptically, which up to now are applied only for UHT-treated milk (Bosch).

   Basically the same products can be filled into plastic bottles and jars as into glass containers. Presently only one system (Bosch) is able to fulfil both necessary demands: that of sterilizing containers from plastics and the filling of particulates. Closing is usually done by heat sealing aluminium lids. For this reason, much attention has to be paid to avoid contamination of heat-sealing rims.
5. Metal cans. As mentioned already only the Dole system is able to apply cans from steel and aluminium for aseptic filling. The existing slit filler, however, limits applications to liquids with very small particles, such as rice.
6. Plastic cans. An aseptic machine for filling and closing of two-piece plastic cans, 'gourmet cans', was recently developed by Shikoku Kakoki. Cans and lids with easy opening feature consist of PP/EVOH/PP. They are sterilized with hydrogen peroxide, UV radiation and heat-sealed inductively. The can is presently offered for liquids only—for example coffee.
7. Composite cans. These may, at present, not be filled with particulate foods, only fruit juice with long fibres.

1.6.2.2 Bulk packaging

Bulk packaging is performed rarely for household use; typical applications are in restaurants and catering for bag-in-box and aseptic transportation from the producer to the food/beverage manufacturer/filler of foods/fruits/fruit purée in bags, which are in crates and drums. The maximum size of particulates is mainly limited by the sterilizing, pumping and metering devices and by the orifice of the filling spouts.

1.7 Typical machines for aseptic packaging of particulate foods.

Some typical machines are discussed in detail.

1.7.1 Form/fill/seal machine for pouches

Figure 1.1 shows the principle of operation of an aseptic vertical form/fill/seal machine for three-sided sealed pouches. The packaging material from reel, usually a complex multilayer material, is sterilized by hydrogen peroxide in a heated bath, which is the syphon lock to a sterile chamber with a slight overpressure of sterile air. In this chamber the film is dried, folded over a shoulder to form a tube and sealed at the long seam. Then the tube, which is closed at the bottom by the cross seal, may be drawn to the non-sterile exterior of the chamber through a tightly fitting flexible lock. Sterile filling inside the chamber is performed using a sine filler. In the tube above, the contents are protected by a neutral atmosphere of sterile

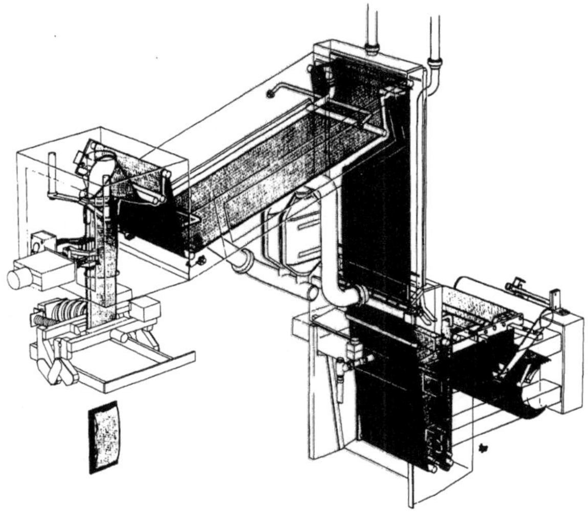

Figure 1.1 Principle of an aseptic vertical form/fill/seal machine for three-sided/sealed pouches for 1-5 l content. Right: film is drawn from reel and sterilized in a hydrogen peroxide bath acting as a syphon closure for the sterile cabinet. Then the film is dried. Left: the dried film is folded over a shoulder to a tube and the long seam sealed. As the tube is closed at the bottom by the cross seal it may be drawn out of the sterile cabinet through a lock, where the grippers for transporting and sealing are located. Filling is performed within the sterile cabinet. (Courtesy of Bosch Packaging Machinery, Waiblingen, Germany).

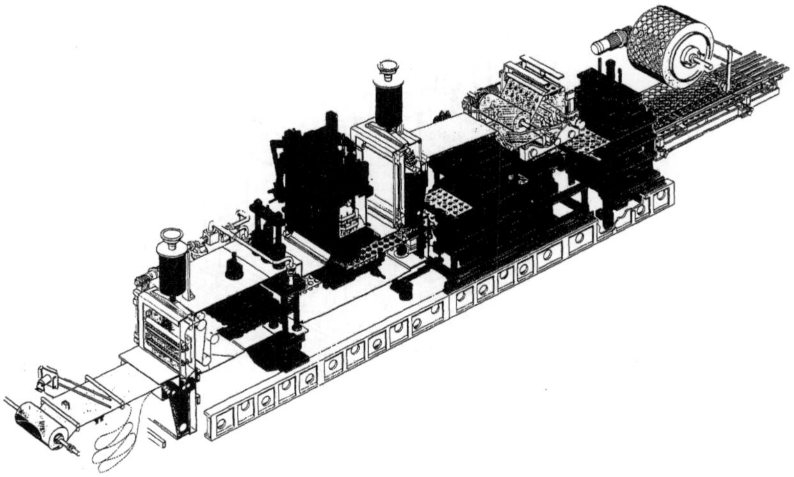

Figure 1.2 Principle of an aseptic thermoform/fill/seal machine. From left to right: (i) unwinding of the cup web from the reel; (ii) splicing table for the web, with box for end of the web; (iii) hydrogen peroxide bath as syphon closure for the sterile cabinet and for sterilizing the web, followed by an air knife to cut off hydrogen peroxide from the web; (iv) idle rollers for the web; (v) local preheating plates for the cup web; (vi) thermoforming station; (vii) filler; (viii) lid film (sterilized and dried in the same way) is put on the filled cup web; (ix) presealing of the webs at the edges creating a tube with closed edges; (x) discharge lock for this tube at the ends of the sterile cabinet; and (xi) outside the sterile cabinet: sealing of the cups, notching, punching, transport of webs and cups, rewinding of waste. (Courtesy of Bosch Packaging Machinery, Waiblingen, Germany).

nitrogen, which maintains a very low oxygen concentration in the headspace of the packs. Grippers spread the sealing zones and vertically reciprocating sealing bars with cutting knifes outside the sterile cabinet are transporting down, sealing and cutting off the pouches. The pack volume is 1–5 l, and output is 15–35 pouches/min, depending on size. Products that are, at present, filled by these machines include various tomato products, sauces such as cheese sauce, and pizza sauce with particulates. Meal constituents and curries could also be filled.

The filling system has CIP and SIP characteristics. Presterilization of the filling system with pressurized steam and of the sterile chamber of the machine by condensed hydrogen peroxide vapours, and also heated air, is performed automatically.

1.7.2 Thermoform/fill/seal machine for cups and trays

Figure 1.2 shows the principle of operation of this type of machine. Both film for the cups and trays and the lid are drawn from reels and transported into the totally closed sterile cabinet through a heated hydrogen peroxide bath. The lower film is heated locally, thermoformed with plug assistance by pressurized sterile air and the formed packs are then filled. Filling is performed by a special piston filler with reciprocating valves having cutting edges (see Fig. 1.3). This filler is able to deposit mixtures with particulates some millimetres in size. The shafts of the sliding valves and pistons penetrate the vessel for the product that has to be filled. At the

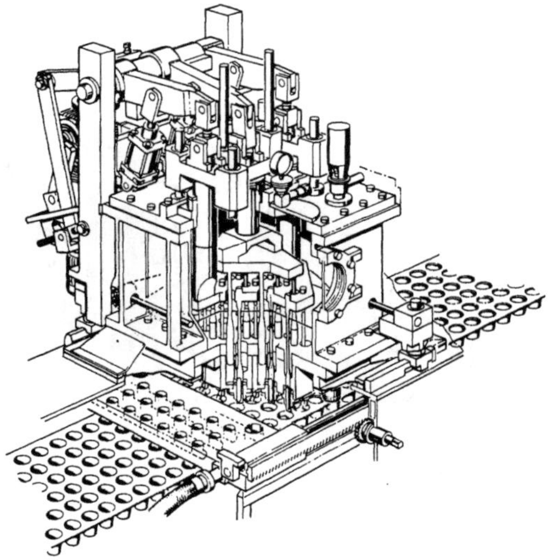

Figure 1.3 Aseptic piston filler, in this case with 24 filling nozzles, pistons and sliding valves. The individual filling heads may be adjusted individually. All adjusted filling heads may be centrally regulated during operation. (Courtesy of Bosch Packaging Machinery, Waiblingen, Germany).

mechanical drive external non-sterile air is separated from the sterile air above the product by rolling diaphragms. For cube sizes up to 12 mm, a filler was developed with a different feed of product to the pistons, filling only one row of packs. This filler was recently modified to fill particulates up to 25 mm cube size. With large particulates, the shape of the containers has to be chosen in a way to ensure uniform distribution of the product. After filling, the lid is applied to the filled web and sealed at the rims. Headspace gas flushing may be performed. Now the webs may

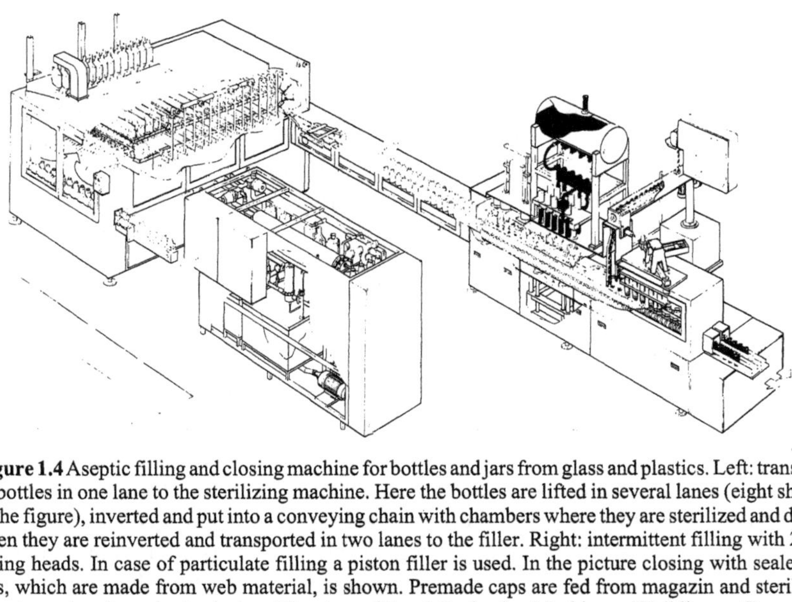

Figure 1.4 Aseptic filling and closing machine for bottles and jars from glass and plastics. Left: transport of bottles in one lane to the sterilizing machine. Here the bottles are lifted in several lanes (eight shown in the figure), inverted and put into a conveying chain with chambers where they are sterilized and dried. Then they are reinverted and transported in two lanes to the filler. Right: intermittent filling with 2 × 5 filling heads. In case of particulate filling a piston filler is used. In the picture closing with sealed Al lids, which are made from web material, is shown. Premade caps are fed from magazin and sterilized in a steam chamber prior to closing. (Courtesy of Bosch Packaging Machinery, Waiblingen, Germany).

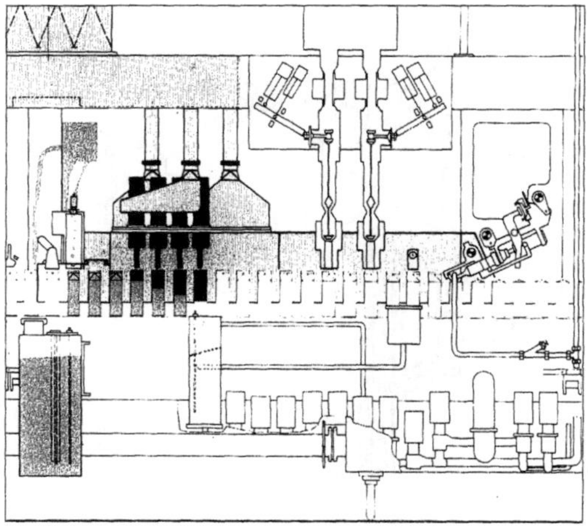

Figure 1.5 Aseptic zone of the Combibloc machine for aseptic filling and closing of cartons from sleeves. From left to right: (i) open cartons with sealed bottom enter the aseptic zone transported by a chain; (ii) hydrogen peroxide vapours are blown into the cartons and then the deposit is dried with hot sterile air; (iii) filling zone with two filling heads; (iv) folding of the top and sealing by ultrasonic sealing heads at the end of the aseptic zone. (Courtesy of Combibloc, PKL, Linnich, Germany).

be transported to the non-sterile outside through a contour lock without risk of infection, where the final sealing of the packs, notching and cutting is carried out.

The CIP of the filling system and presterilization of this system and the aseptic cabinet are similar to the previously mentioned pouch machine.

1.7.3 Filling and closing line for bottles and jars

Figure 1.4 shows the principle of a filling line for bottles and jars. The containers, which are precleaned and heated by a special rinser, enter the sterilizing machine in one lane and they are sterilized in several lanes upside down, by treating inside and outside with hydrogen peroxide vapours and then drying with sterile air. The containers are inverted and transported intermittently to a piston filler which is capable of filling particulates with a cube size of 12 mm as a maximum. This filler has the same basic design as used in the thermoform/fill/seal machine. In the next machine, also operating intermittently at sealing, the containers are closed with metal caps, which were sterilized with pressurized steam in an UHT process when entering the machine. For liquid products, magnetic-inductive metering devices are used for filling. For plastic bottles heat-seal closures from foil are applied.

CIP and SIP for the filling system and sterilization of the closed aseptic cabinet are carried out in the same way as indicated for the other machines. The output of the machine is 6000 units/hr.

1.7.4 Form/fill/seal machine for cartons from sleeves

In the Combibloc system premade sleeves (folding boxes) are fed from a stack to a turret with mandrels, in which the bottom of the pack is sealed (finseal with gussets) and folded. The packaging material for aseptic purposes consists (from

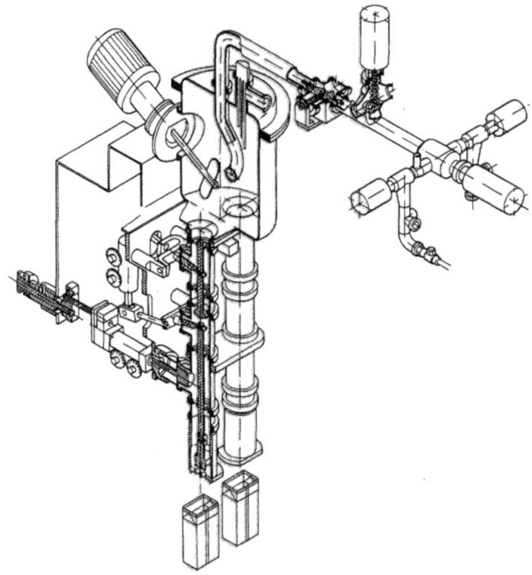

Figure 1.6 Particulate filler with pneumatically operated piston and valves and a vessel with an agitator for even distribution of particles and liquid. (Courtesy of Combibloc-PKL, Linnich, Germany).

inside to outside) PE/Al foil/PE/carton/print/PE. The long seam of the pack is split and folded back before sealing. After bottom-sealing, the packs are erected and pushed into a two- or four-line transporting chain.

In the transporting chain they are sterilized by treating with a mixture of heated air and vapourized hydrogen peroxide with subsequent drying by heated sterile air (Figure 1.5). Filling may be performed in one or two goes, while for liquids a magnetic-inductive metering system is used. For particulates a piston filler is used with a buffer tank and paddle stirrer (Figure 1.6). The valves are specially adapted to the characteristics of the product. The maximum particle size is 20 mm, the maximum percentage of particles 50% and the viscosity range 80–10 000 mPs (Ostermann, 1990). Sealing of the top is effected ultrasonically. Steam or sterile nitrogen may be injected into the headspace prior to sealing.

Presterilization of the relevant machine areas is also performed using hydrogen peroxide vapours and hot air followed by drying with sterile hot air; pressurized steam is used for presterilization of the filler and the filling lines. The aseptic zone of the machine is vented with sterile air from a vertical laminar flow. Typical packaged particulate foods are soups and sauces, tomato products and fruit dessert. The output is 5000 cartons/h.

1.7.5 Form/fill/seal machine for cartons from the reel

For filling high-viscosity products or liquids/pastes with particulates in the Tetra Pak system the TBA machine with a deep bath of heated hydrogen peroxide is used. The tube of packaging material, in which filling is done, is manufactured in a conventional way. The machine for mixtures with particulates uses the headspace principle. While conventional machines (for instance for packages with milk and fruit juice without headspace) maintain a constant level of liquid within the tube and the cross seals are made by pressing off the liquid from the sealing area by the sealing jaws, with the headspace principle liquid is constantly metered into the package and pressing and sealing of the cross seals is performed before the package is completely filled. For filling liquid products a source of pressure and a cone valve for regulating the flow are utilized. High-viscosity liquids are pumped into the package by a lobe pump. For particulate mixtures, liquid in one side and a particulate mixture with little liquid in the other are dosed separately. Liquid is fed continuously through a central pipe and particulates discontinuously, in order to prevent particles adhering in the sealing zone, through a concentric external filling pipe. Maximum particle size is presently a 10 mm cube. The particulates are transferred by a piston pump with a rotating cone which acts as a valve for sucking and dispensing. While the percentage of particulates is currently limited, machines for the future are planned to fill any ratio of liquid to particulates. The headspace may be filled with neutral gas.

Filling lines and pumps are sterilized with pressurized steam, external surfaces of the pipes and the sterile air/gas ducts with hydrogen peroxide. A precondition for particulate filling is two separate streams of product; for instance from two tanks. This may be achieved either by separating the concentrated mixture of particulates and the liquid after sterilization, if sterilized in one unit, or by separate sterilization of the two streams of product.

1.8 Conclusion

Aseptic packaging of food and beverages generally has gained greatly in economic importance, mainly for milk and fruit juices. Its application is spreading—especially in areas with inadequate refrigeration options. Packaging particulates in liquids and pastes into consumer packs is still a very young field of application with only a short track record. The technical options for the two bottlenecks—sterilizing and filling—have developed to a commercial level of application and continue to be developed, improved and adapted to the special requirements. The aim of this contribution was to provide an overview of the state-of-the-art and to give stimuli for further improvements and applications.

References

Anon. (1981) *Neue Verpackung*, **1:** 53.

Buchner, N. (1989) Anlagen zur aseptischen Befüllung von Flaschen aus Glas und Kunststoff, *Die Molkerei-Zeitung-Welt der Milch*, **43(3):** 84–87.

Buchner, N. (1991) Reinraumtechnik in Maschinen für die Sterilabpackung von Lebensmitteln und Getränken, *VDI-Bericht*, **919:** 263–282.

Buchwald, B. (1987) Anlagen zur schonenden UHT-Behandlung von flüssigen Lebensmitteln mit festen Partikeln. In Reuter, W., (ed.) *Aseptisches Verpacken von Lebensmitteln*, Behr's Verlag, Hamburg, pp. 60–72.

Cerf, O., and Hermier, J. (1972) Diversité de la résistance de spore de bacillus a l'eau oxygenée, *Le Lait*, **52:** 1–20.

Denny, C. B., Brown, C. K., and Yao, M. G. (1974) NCA-tests of Primepak aseptic canning machine using hydrogen peroxide for equipment and container sterilization, Research Report No.1, Washington, Research Laboratory of the National Canners Association.

Emch, F., (1978) Aseptische Verfahren in der Lebensmittelindustrie, *Lebensmitteltechnik*, **11:** 5–21.

Emch, F., (1991) Lebensmittel aseptisch verpacken, *Lebensmitteltechnologie*, **24:** 118–126.

Hastings, A. P. M., and Murray, S. A. (1987) Pre-sterilization of Particulate Foodstuffs with a Liquid Phase, In *Aseptisches Verpacken von Lebensmitteln*, Reuter, W., (ed.), Behr's Verlag, Hamburg, pp. 82–90.

Haubs, H. (1988) Die Entkeimungsfiltration in der Getränkeindustrie, *Pharma-Technologie Journal*, **3:** 15–17.

Hayashi, R. (1989) Application of high pressure to food processing and preservation: philosophy and progressing, ICEF 5, Cologne, 28 May–3 June, Abstract No. 5.09.

Hermans, W. F. (1988) 'In-flow' fraction specific thermal processing (FSTP) of liquid containing particulates, Internat. Fachtagung 'Sterile Prozeßtechnik', 7–8 November, Frankfurt/M., pp. 54–61.

Iwamoto, J. (1985) Heating sterilization and food packaging by microwave, *JPI Journal*, **22:** 48–52.

Kessler, H. G. and Fiedler, J. (1980) Zur Frage der kontinuierlichen UHT-Durchflußsterilisation von Flüssigkeiten mit stückigen Teilchen, *Zeitschrift für Lebensmitteltechnologie*, **31:** 124–127.

Khaladadij-Nia, J. und List, D. (1981) Energieverbrauch bei der Herstellung sterilisierter tiefgefrorener und getrockneter Gemüseprodukte, *Zeitschrift für Lebensmittel-Technologie,* **1**: 14.

Loire, P. (1976) Aseptic packaging of food products in drums, IAPRI Conference, Munich, 9–11 June.

Lubieniecki, M. (1973) Bedeutung des Keimgehalts von Packstoffen Gezw. Packmitteln für nicht-sterile Lebensmittel, *Verpackungs-Rundschau*, **24**: 77–84.

Mankowski, A. (1987) Innovationen in der Lebensmittelindustrie durch die aseptische Kartonpackung, *Verpackungs-Rundschau*, **4:** 356–360.

Meredith, R. J. (1988) Principles and Relevance of Microwave Processing of Food with Particles, Internat. Fachtagung 'Sterile Prozeßtechnik', 7–8 November, Frankfurt/M., pp. 87–95.

Ohlsson, Th. (1988) Continuous Heat-exchanger System for Processing of Pumpable and Particulate Foods, Internat. Fachtagung 'Sterile Prozeßtechnik', 7–8 November, Frankfurt/M., pp. 9–24.

Ostermann, A. E. (1990) *Aseptic Carton Packages From Sleeves*, Behr's Seminare, Cologne.

Placek, L. and Witter, K. Th. (1972) Bakteriologische Oberflächenuntersuchungen von Einmalpackungen, *Verpackungs-Rundschau*, **23:** 742–748.

Reitler, W. (1990) Aufheizverhalten stückiger Lebensmittel bei Ohm'scher Widerstands-Erwärmung im Vergleich mit indirekten Erwärmungstechniken, Diskussionstagung 'Keimarmes und aseptisches Verarbeiten von Lebensmitteln', GVC-VDI-Ges. Verfahrenstechnik, Cologne, 27–28 November, Preprints, pp. 41–58.

Schubert, H. (1990) Grundlagen und Besonderheiten der Mikrowellenerwärmung von Lebensmitteln, GVC-Tagung Keimarmes und aseptisches Verarbeiten von Lebensmitteln, Cologne, 27–28 November, Preprints, pp. 59–79.

Sitzmann, W. and Münch E. W. (1990) Betriebs- und produktgerechte prozeßintegrierte Keimreduzierung, Diskussionstagung 'Keimarmes und aseptisches Verarbeiten von Lebensmitteln', GVC-VDI-Ges. Verfahrenstechnik, Cologne, 27–28 November, Preprints, pp. 127–143.

Skudder, P. J. (1988) Practical Relevance of the Ohmic Heater for the Sterilisation of Particulate Food, Internat. Fachtagung 'Sterile Prozeßtechnik', 7–8 November, Frankfurt/M., pp. 76–86.

Streubl (1988) Der mikrobiologische Zustand von Glasverpackungen bei ihrer Erzeugung und Lagerung, Diplomarbeit Universität Hohenheim.

Sturm, W. and Gilliand, A. (1974) Einsatz physikalischer Sterilisationsverfahren bei der Herstellung aseptischer Packungen, *Verpackungs-Rundschau*, **25**(4): 298–302.

Swartling, P. and Lindgreen, B. (1962) *Aseptic filling in Tetrapak—Sterilization of the Paper. Milk and Dairy Research*, Alnarp Report No. 66.

Thum, J. (1982) Entkeimung von Packstoffen für die Sterilabfüllung von sauren Lebensmitteln, Diplomarbeit Universität Hohenheim.

Vögele, P. (1992) *Aseptisches Verpacken hochviskoser und stückiger Lebensmittel*, GDL-Symposium "Fertiggerichte - Fast-Food Catering", Stuttgart-Hohenheim, 13 March.

Voss, E., and Moltzen B. (1973) *Milchwissenschaft*, **28:** 479–484.

Wallhäuser, K. H. (1988) Die Entkeimungsfiltration im pharmazeutischen Bereich, *Pharma-Technologie Journal*, **3:** 8–15.

Wernimont, D. V. (1988) Latest Developments in Continuous Processing of Viscous Particulate Food Products, Internat. Fachtagung 'Sterile Prozeßtechnik', 7–8 November, Frankfurt/M., pp. 44–53.

Witter, K–Th. (1979) Mikrobiologische Bedingungen von kunststoff- oder wachsbeschichteten Papieren für die Lebensmittelverpackung, *Coating*, **6**:.

2 The market for aseptic products, processing and packaging systems

E. M. A. WILLHOFT

2.1 Introduction

Packaging is in a continuous state of flux, constantly prodded along by developments in materials, packaging machinery, the products it is designed to protect, and by the purchaser and consumer. Not long ago microwave technology represented a concept of technical curiosity when applied to heating food and a technique dabbled with by a minute section of the industry. Current estimates in decreasing order for the USA, Japan, Australia, the UK and Canada indicate a household ownership of microwave ovens ranging between 70% and 40%. The goalposts on acceptable materials coming into contact with foods now need to be shifted once again to allow for the possibility of thermally enhanced migration of trace chemicals to food from the packaging during heating with microwave energy.

Packaging developments are now in their most exciting phase, having to meet the challenges throughout the advanced countries of consumer demands of increased convenience, preoccupations with health and diet foods, aseptic preparation and the use of lighter unit packs involving paper, composites and thermoform types. In addition, social changes affecting eating habits are taking place at an unprecedented rate, providing the motivation for continual innovation in packaging and food processing.

Legislation and levies have led to a rapid growth in recycling domestic waste, particularly in continental Europe. The amount of unrecycled solid waste in Europe varies between 86.5% (Austria) and 95.7% (Greece). European legislation is targeting a recycling level of 60% by the mid 1990s and 90% by the year 2000, which will exert an enormous influence in structuring the direction taken in selection of materials in packaging into the next century. At present, the international league table in recycling reflects levels of governmental support rather than any voluntary pattern of behaviour. Austria, Spain and the Netherlands head the list in this order, while Greece, the UK and France trail the other EC countries, with Greece at the bottom of the league, clearly reflecting that it does not yet pay commercially to recycle in these latter countries. A recent report from Environmental Policy Consultants claims the number of major companies in the UK that have adopted environmental policies has risen from 41% to 66% during 1991. However, 74% of companies said they wanted the government to help by enforcing legislation more effectively (Environmental Policy Consultants, 1992). The recent British Environmental Protection Act 1990, which came into force in April 1991, emphasizes the benefits of recycling waste and saving energy. In the USA, only 11% of waste is recycled compared with 50% in Japan; however, the Japanese are probably the highest users of packaging on a *per capita* basis. In 1987, the *per capita* expenditure by Japan was £217 compared with £96 in the UK.

The attitude to recycling in the USA and the UK has taken a pounding from the environmentalists due to retailers making confusing and misleading claims on consumer goods, which has prompted the need to define in *quasi*-legal form such terms as: (i) recycled; (ii) recyclable; (iii) degradable (including biodegradable and photodegradable); and (iv) compostable.

The environmentalists have to be given some of the credit for highlighting the need to design packaging with a cradle-to-grave philosophy. Retailers and consumers benefit by getting more reliable information for selecting materials having least impact on the environment, while companies become increasingly pressurized into adopting processes and materials that are conveniently disposed of in an environmentally friendly way. A consequence of new European legislation may well be that export of certain products will be blocked on the grounds that the exporting country may not be conforming with the set guidelines. The European Court of Justice has already ruled that, in some cases, individual members states can stipulate packaging requirements rather like returnable bottles, when the benefit is environmental. The promised abolition of frontier controls within Europe creates opportunities for trade as well as the potential for new health hazards.

Food technologists need to keep pace with new processes and materials, packaging technologies and consumer education. There is a view that microorganisms are constantly evolving and adapting to new hosts, as well as developing resistance to conventional microbial inhibitors and preservatives. Certainly, there is evidence to support this theory from medical science. Packaging is now capable of being designed from basic scientific principles and, therefore, the need to carry out exhaustive trials has become less vital.

About 30% of the world's food is wasted post-harvest (*cf.* 11% waste in the USA) due to inadequate protection and storage. With the demise of communism and its replacement with capitalism, there is clearly considerable scope for reducing waste to a substantially lower level by the exploitation of the technologies taken for granted in the advanced countries; that is, in aseptic processing and packaging including gas blanketing, irradiation, refrigeration and so on.

Originally, irradiation had to be labelled as if it were an additive. In the USA, it is now accepted as a process akin to heating and freezing. The increasing costs of refrigeration energy and chilled storage space, coupled with the tightened legislation on permissible temperature fluctuations, are stimulating growth of ambient shelf-stable products manufactured by aseptic processing and packaging techniques. Separate sterilization of the edibles and the packaging is central to the definition of aseptic processing and packaging, in contrast to retorting which provides the sterile environment post-packaging by *in situ* heating with the attendant problem of over-cooking to ensure total sterility within the pack. Aseptic foods are classified into three chemical environments:

1. Low acid with pH above 4.6
2. Acid with pH of 4–4.6
3. High acid with pH of less than 4.0.

As a result of the relationship between microbial viability and pH, different thermal conditions are applied to the different categories during processing. High-acid foods require lower temperatures and holding times. Conversely, low-acid foods require higher temperatures and longer holding times.

Until the early 1980s the USA were behind the Europeans in aseptic processing, simply because it was only in 1981 that the FDA gave approval to the use of hydrogen peroxide (H_2O_2) as a packaging sterilizer—even then it was limited to polythene contact only. However, the FDA followed established European practice to include other materials in 1984. Prior to 1981 the sole method of sterilization was with steam. The most common packaging sterilant used now is H_2O_2, generally in combination with heat. Sterilization is effected by direct steam injection or, indirectly, by using one of three types of heat exchangers: plate, tubular and scraped-surface exchangers. Fuller descriptions of the different modes of effecting sterilization are covered in other chapters (1, 4–9); suffice it to say that the type of product (pH, viscosity of liquid phase and size of particulates) determines the type of plant used in the aseptic process.

Low-acid soups and sauces containing particulates are now a commercial reality in France and Germany but still account for less than 1% of all aseptically packed products. Apart from identifying an enormous growth potential for particulates, the embryonic state of this sector of the market makes it an unreliable indicator of its true growth potential. Soft fruits of up to 25 mm in diameter have been successfully sterilized in scraped-surface heat exchangers as well as low-acid vegetable products of 15 mm size. Processing of particulates through the scraped-surface exchanger reaches a limit of tolerance at about 15 mm, simply because of overcooking. The liquid phase needs to have a high viscosity to prevent separation and, for this reason, other methods of processing are being and have been developed by APV, while Alfa Laval and Stork base their process on the separate heating of the liquid and particulates. CMB and Bosch are also active in the field of particulates, the former on the filler side. A liquid phase can be added through the filler cut-off head forming a dilution of products such as soup, or to add a garnish to the top of a curry meal or dessert. Thus, a layered fill of individually processed product including particulates is produced. Robert Bosch Packaging Machinery have developed an aseptic form, fill and seal machine using high-barrier, microwaveable material. A move towards microwaveable cartons is also claimed by Bowaters by eliminating the aluminium foil layer in the preformed cartons and replacing it with EVOH, which is a barrier to oxygen.

Ohmic or resistive heating, developed commercially by APV, must be regarded as a major development, along with microwave heating developed by Toppan of Japan and Berstorff Bio-Technik of Germany. The first commercial bulk packaging of 20 mm vegetable particulates for the catering market in the UK started in autumn of 1989 using ohmic heating. Particulates up to 25 mm with 12 months' shelf-life are currently produced by the ohmic (APV) process, which is about twice as large as can be produced on the Combibloc system, which led the technology in particulates prior to APV's entrance. Stork of Amsterdam operate a pilot plant facility with particulates but are lacking in evidence of any commercial installation.

The challenge of low-acid products containing particulates, coerced by consumer demand, has driven processors to start manufacturing such products on both sides of the Atlantic. Frost and Sullivan (1991) conclude that such products will become a commercial reality in the mid-1990s in the USA. Indeed, they state that particulate, high-acid products with particles up to 50 mm in size have already been successfully processed and packaged in bulk for the institutional market using large bags enclosed in a box or drum.

On both sides of the Atlantic the range of products that are aseptically packaged includes the following:

- Milk
- Fruit juices
- Condensed milk
- Chocolate puddings
- Edible oils, including butter and margarine based on rapeseed
- Tomato products
- Tofin (Japanese soyabean curdblocks)
- Soups and gravies
- Scrambled eggs
- Jelly and coffee puddings
- Fruit salads

Aseptic packaging is understood as the bringing together of presterilized edible products and sterilized packaging and filling the product under sterile conditions. This approach to producing an aseptic pack is to be contrasted with that based on filling a container, sealing it and subjecting the whole pack to heat sterilization, as happens in retorting where, to ensure complete sterilization throughout the pack, a degree of over-cooking is inevitable. Aseptic processing can thus be considered to produce an organoleptically superior product. The main benefits that will guarantee continued growth of aseptic technology are: (i) improved product quality; (ii) lower costs; and (iii) added convenience.

Hot filling is also adopted as a way of sterilizing the pack, but is not considered under the heading of aseptic processing and packaging. Both retorting and hot filling produce a sterile pack that suffers from a degree of nutritional loss due to the length of time that the product is kept hot. In aseptic processing, the edibles are subjected to flash heating and cooling, which sterilizes product without loss of nutritional value. Heating is effected either with one of a variety of heat exchangers or by direct injection of steam into the product, or more recently by resistive heating of the edible products. It has been suggested that maintaining sterility by the use of additives has prevented infections by food-poisoning pathogens such as

Salmonella spp. (Wheelock, 1988) and the fashion for removing additives is partly responsible for the rapid rise in *Salmonella* infection. Wheelock cites research conducted a few years ago, which concluded that the hazards associated with microbial contamination of food exceeded those linked to additives by a factor of about 100 000, yet the public perceived the risk associated with food additives as high, while there is a small concern for food-borne disease. When sterilizing the packaging, the various suppliers of plant use a variety of processes and chemicals, either singly or in combination, as follows:

- Superheated steam
- Chlorinated water
- Hydrogen peroxide
- Dry hot air
- Hot water with or without citric acid
- UV radiation, with or without hydrogen peroxide, and/or acetic acid
- Ionizing radiation

Recently, the technology of stripping off an adhering film to produce a sterile packaging surface has been employed successfully. Each supplier tends to adopt one type of system to ensure full sterilization of plant and packaging. In the case of plastic containers (e.g. in a form/fill/seal process) the heat generated in the blow-moulding process provides the high bacteriological standards required to ensure full pack sterility.

2.2 The west European market

Market data for Europe has been supplied by Leon R. Frotscher Associates (1987) coinciding with the survey carried out during April to September 1987. The countries included in the survey were UK, Germany, France and Italy, with Belgium, Netherlands, Sweden, Finland, Switzerland and Spain having a lower input.

There are now in excess of 1600 and probably 2000 aseptic packaging systems operating in western Europe, cartons accounting for over 85% of all systems installed, of which 80–85% are Tetra Pak, the remainder being PKL Combibloc. Combibloc have successively developed particulate filling using their preformed cartons. Tetra Pak still rely on sealing their cartons through the fill, which limits their ability to handle particulates. Preformed and form/fill/seal plastic pots account for over half of the remaining installations supplied by Bosch, Gasti, Hamba, Ampack, Metal Box and Benco. Plastic bottle installations are supplied by Remy, Rommelag and Serac, while bag-in-box systems are offered by Bowater, Coloreed, Oettel and Elpo.

Cartons for filling liquids were a major trend in the 1980s, but the growth became somewhat restrained by poor opening and closing of the packs. Further problems occurred from the environmentalists who claimed that the mixing of materials prevented reclamation. The carton manufacturers have therefore taken to funding their own development programmes. Bearing in mind that 100 million cartons for, say juice, produce nearly 2 900 tonnes of packaging waste, Tetra Pak in 1987 devised ways of re-using factory waste to make chairs, briefcases and even building materials, in the hope of removing any marketing impediment to the growth of their aluminium-laminated cartons.

Of the 1600 installations in Europe in 1986, Table 2.1 gives a breakdown by pack type.

Prepac and Thimmonier supply pouch-filling systems that are restricted to only 15–20 days' shelf-life for milk. Systems for filling metallized pouches are supplied by Elecster. The annual market for aseptic systems is about US$100 million with each system costing between $0.4 million and $1.6 million. About 150–200 new plants are installed each year, about one-half being straight replacements and the rest new installations. The proportion of new systems dedicated to cartons (*c.* 75%) is expected to fall slightly over the next few years as more emphasis is directed at thermoformed plastic containers, plastic and glass bottles, and bag-in-box systems. The high level of acceptance of the aseptic bricks dominated by Tetra Pak and Combibloc will ensure that their method of packaging is favoured for many new applications. New systems are being developed by SIG, as well as a joint venture involving CMB and Elopak.

According to Frotscher Associates, in 1986 around 13 billion litres of product were filled into an estimated 20 billion containers. In 1991, this was projected to increase to 16 billion litres (26 billion containers), which assumed an annual growth of 4.5% in volume packed but nearer to 5.5% in unit terms, indicating a proportionally higher value of packaging by 1991. Cartons accounted in 1987 for about 90% by volume of all aseptically packed products of which more than 95% were for milk and fruit juices. Unless Tetra Pak bring in their 'much-hinted-at' technology on particulate filling they are likely to lose market share to Combibloc who have recently produced a 2-l cartoning system. Frotscher Associates estimate a decline in the carton sector from 90% in 1986 to 88% in 1991. Other developments in cartons are resulting in new shapes to avoid the commodity image of the brick

Table 2.1 Installation of aseptic packaging plant in western Europe (after Leon R. Frotscher Associates, 1987)

Type of pack	Operational installations	
	Number	%
Cartons	1400	88
Thermoformed plastic cups	100	6
Plastic bottles	28	2
Cans		
Metal	13	1
Laminated	1	<1
Aerosols	6	<1
Satchets/pouches	8	1
Bag-in-box (<20 l)	35	2
Total	1591	100

Table 2.2 Volumes packed (billion litres) in different pack types (after Leon Frotscher Associates, 1987)

	1986	1991
Cartons	12	14
Plastic cups	0.4	0.6
Plastic bottles	0.4	0.6

shape and to enable the filler operator to single out his product in the market place by shape. Nearly 4 billion thermoformed plastic containers were aseptically filled in 1986 although, according to Frotscher Associates, they accounted for only 3% of the volume of product packed, of which nearly half were single-portion (14 ml) cups for caterers for milk, cream and coffee cream. Other were mainly 125 g cups for yoghurt, desserts and cream. Table 2.2 gives the projected growths up to 1991 for cartons, plastic cups and plastic bottles.

Excluding the mini-portion packs, the market in plastic containers is expected to grow rapidly—especially in desserts and fruit compotes. Many dairy products are limited to a 3 months' shelf-life because of inadequate barrier properties of the thermoformed plastics (polypropylene or polystyrene) to oxygen. However, such non-laminated plastics favour microwave reheating, which is an important influence in the current growth pattern. Serac and Remy systems are highly regarded, which in turn will promote use of this equipment for plastic bottles as an alternative pack to the carton, mainly in the lower added-value markets such as milk, fruit drinks, cream and flavoured milk. Aerosols have gained a substantial share of the cream and topping markets in the UK and Belgium, but not in Germany because of the alleged environmental damage caused by them. Sachets and pouches are limited to a few installations filling UHT milk in Spain (with a 15-day shelf-life) and producing sterile mix for ice cream in Scandinavia. Prospects for retail pouches are generally poor for Europe, although the outlook for the catering sector may be better either for bulk or single-portion sauce packs.

Bowater, Coloreed and Oettel all market low-acid products in bag-in-the-box systems and these systems are well established for fruit and tomato products. Non-dairy cream, ice-cream mix and milk shakes are now filled in this manner, although the future market will depend to large part on the acceptance by the catering trade of the bag-in-box concept as it would compete directly with cartons, pouches and plastic bottles. Bosch, Serac and others are promoting aseptic systems for filling glass containers in Germany, which could open the route for expanding glass packaging. Other developments will exploit shape differences, and opening and handling features designed to provide presentation advantages over standard cartons as, for example, with the Bosch Hyperpack. According to Frotscher Associates, of the 13 billion litres of liquid and viscous products packed in 1986, over 90% was accounted for by milk and fruit juice; however, the division between the two product types varied greatly from country to country as do the projections for future growth. An analysis of aseptic product types in western Europe is summarized in Table 2.3.

Table 2.3 The European market for aseptic packaging by type of product (after Leon Frotscher Associates, 1987)

Product	Volume (Ml) 1986	1991	% Change p.a.
Milk/buttermilk	8471	9731	2
Fruit juice/fruit drink	3150	4296	6
Flavoured milk/milk drinks	394	541	6
Evaporated/condensed milk	25	48	14
Coffee milk and creamer	62	84	5
Cream	151	202	5
Yoghurt	210	251	3
Ice cream/milk shake	44	57	4
Non-dairy cream and topping	25	33	5
Desserts	140	200	8
Fruit filling/compote	25	58	20+
Tomato products	79	126	10
Fruit	—	0–20	—
Vegetables (excluding tomatoes)	—	0–20	—
Soup	12	148	20+
Sauces	5	40	20+
Ready meals	—	0–10	—
Baby food/baby milk	10	20	14
Pet food	—	0–20	—
Other	40	80	—
Total	12843	15950	—

Frotscher Associates conclude that, in Europe, aseptic processing and packaging were being driven by technology rather than by the market. Technical innovation and development within companies in the late 1980s achieved its own impetus, with little relation to the perceived requirements of the market. However, legislation and commercial incentives in recycling are influencing packaging among the more environmentally conscious countries.

2.3 The US market

Data on the American market in systems and products are based on a comprehensive report produced by Frost and Sullivan in the summer of 1991, which is recommended for further reading. The principal suppliers of aseptic processing systems are considered in turn, with a summary of the principal features of each system.

2.3.1 Alfa Laval

Alfa Laval produce a range of systems to cover both high- and low-acid foods in viscous and particular form. They include:

- Contherm
- Vacu-Therm Instant Sterilizer (VTIS)
- Steritherm

- Viscotherm
- Spiraflo

The Contherm is a scraped-surface heat exchanger that handles a wide range of pumpable products that are viscous, sticky or lumpy; for example, tomato paste, cheese sauces, edible oils, fruit sauces and pulp, peanut butter, malted milk, yoghurt, margarine, shortening, and so on. Temperatures of the heat-transfer fluid can vary from 230°C to –60°C. The Contherm heat exchanger is designed for vertical positioning and, therefore, this is of benefit where floor area is at a premium.

The VTIS is a UHT system comprising a balance tank, a plate heat exchanger, steam injector, flash cooler, homogenizer, pumps, valves and control panel. It injects steam directly followed by flash cooling in an expansion chamber equipped with a condenser. Product is preheated to about 80°C with water injected with steam in the heat exchanger. Direct steam injection raises the temperature to 140°C for 4 s, followed by flash cooling to about 80°C in an expansion chamber fitted with a condenser. The product then is homogenized, cooled to 20°C and is ready for storage in an aseptic container or fed directly to the filler. The VTIS is capable of processing milk, flavoured drinks, cream, puddings, soy-based drinks, juices, soups and sauces, and so on. Alfa Laval claim that the plant can be switched from one product to another without a break in production.

The Steritherm is also a UHT system but features indirect heating in a plate heat exchanger. It is designed for low- and high-acid foods. Its principal strength is the low cost of operation compared with the VTIS and claim savings of 60% of the total operating cost of typical direct UHT systems, with approximately 85% savings stemming from steam and water. The Steritherm has largely superseded the more costly VTIS system.

Viscotherm is designed for high-viscosity products and is suitable for juices, diced fruits, jams, puddings, vegetable purées, sauces, soups, and so on. Several pumps with varying handling characteristics may be installed to fit in with end-product requirements. Product is heated with slow and gentle steam heat, and then passes on to a second Contherm heat exchanger to arrive at the desired sterilization temperature of about 140°C. Product is then moved into a holding tube for a set period and cooled to the filling temperature in three stages by flowing the product through three Contherm coolers. Homogenization facilities are available in the system, including an aseptic intermediate storage facility.

The Spiraflow embraces four types of corrugated tube-type heat exchangers, producing a family of units. The particle size and viscosity of the continuous phase determine the type of heat exchanger required. Products range from milk through to yoghurt and fruit pulp with large particulates, and so on.

2.3.2 APV Crepaco

APV supply variants of the complete aseptic processing system in heat exchangers, homogenizers and uperizers. Each is briefly considered.

2.3.2.1 APV JuiceMatic Aseptic System. The JuiceMatic system operates with indirect heating and is designed to process a wide variety of juices, juice-based drinks and juice concentrates in long continuous runs. It uses the 'Paraflow' plate heat exchanger and is controlled by APV's 'ACCOS 2s' computerized system to various levels of automation. A de-aerator is included to minimize the effects of oxygen on the product. The system operates by sterilizing first with hot water and then cooling it. Heated product moves to the holding coil where it is sterilized and then cooled in the heat exchanger. Aseptic product can be stored until required for filling.

2.3.2.2 APV Super Ultramatic Aseptic System. This unit is intended for low-acid foods ranging from milk and cream to toppings, puddings, snack dips and sauces, and provides homogenization facilities. As with all APV aseptic processing systems, it is fully compatible with all modern filling systems. With the exception of product homogenization, higher temperatures and shorter holding times, the Super Ultramatic system is similar to the JuiceMatic unit. In operation, homogenized product is rapidly brought to 285°C and held for less than 3 s before being cooled by regeneration to minimize steam and water usage.

2.3.2.3 APV Uperizer Aseptic System. The uperizer is designed to handle low-acid foods such as milk, cream, puddings, toppings, dips and sauces and even more viscous material. It operates by direct steam injection and preheats raw product, which is then fed into the uperizer in which the steam is injected, raising the product temperature to 150°C. The holding time is 2.5 s after which the injected steam is flashed off to bring the product down immediately. As with all APV systems, the Uperizer can be used with the 'ACCOS 2s' series of microprocessors.

2.3.2.4 APV Crepaco Heat Exchangers. APV Crepaco supplies five different types of heat exchangers to suit juices, drink and concentrates, to more viscous products such as sauces, purées and gravies, and also products containing large particulates. They are divided into indirect and direct heating/cooling units. The indirect heat exchangers include:

1. Plate heat exchanger, for example the Paraflow.
2. Scraped-surface heat exchanger, either horizontal or vertical mounting.
3. Tubular heat exchanger available in three different types: the Paracel; the Tube-in-Tube (TNT); and the Coiled Tubular Aseptic (CTA).

The direct heat exchangers include:

1. Direct steam injection system.
2. Infusion heating system.

2.3.3 ASTEC

ASTEC (Aseptic Technology Engineering Co.) provide a complete system ranging across both processing and packaging for the food, dairy and pharmaceutical industries. They supply a full turn-key service and license their technology to the USA and other countries. Specific plant that they supply includes:

1. Heat exchangers (tubular, scraped-surface, plate, steam-injection and infusion).
2. Sterilizing and pressurizing systems.
3. Product handling and storage plant (pumps, tanks, vessels, drainers).
4. Specialized equipment (ASTEC aseptic bin, ASTEC aseptic bag unloader, ASTEC raque depositor for particulates).
5. Accessories (piping, valves, metering devices, etc.).

The ASTEC HydroCoil heat exchanger is an indirect, tubular design with a spiral configuration having the heat exchanger located in a coil. The turbulence created by the coil geometry facilitates heat transfer by moving particles along the tube wall. Product temperature control and energy regeneration are provided by means of recirculating water. The heat-exchange tube is available in sizes ranging from 6.35 to 76.2 mm and is capable of handling up to 32 760 l/h, depending on the type of material passing through the sterilizer. The HydroCoil is used by the major food packers in the USA including Borden, Dean Foods, Hunt-Wesson, Knouse Foods, Land O'Lakes, Real Fresh and others. It is used for the processing of juices, drinks, milk and cream, to viscous products (such as apple sauce, cheese sauce and puddings) to particulate-containing products such as soups, fruit toppings and fruit suspensions for yoghurt and ice cream.

The ASTEC Aseptic Drainer is a pressure vessel that is equipped with a rotating or vibrating screen, which separates the liquid from particulates. It is used for the purpose of eliminating poor distribution of particulates in the liquid. When products containing particulates are processed, the ASTEC Aseptic Drainer may be used upstream from the cooler in order to ensure that the particulates making up the final product are in a prespecified proportion to the water, brine or syrup.

2.3.4 Cherry Burrell

Cherry Burrell market the Unitherm, a tubular aseptic processing system, which is pre-assembled on a single frame, prepiped and wired, complete with valves, pumps and necessary controls and ready for operation. It is equipped with the Spiratherm tubular heat exchanger, a homogenizer and a de-aeration unit, which is intended to remove oxygen and off-flavours. The Spiratherm heat exchanger has a trombone-type spiral product channel that prevents hot spots forming by virtue of rapid linear speed of travel, which also prevents the build-up of debris. The Unitherm is best suited to low-acid products such as milk, cream, toppings and infant products. However, it can be adapted to take high-acid products. The Thermaflex is similar in design to the Unitherm system except for the use of a plate heat exchanger instead of the tubular design with the latter. It is best suited for high-volume processing of single-strength juices, juice drinks and beverages. By installing a Spiratherm, or other tubular heat exchanger, viscous and pulped products are readily handled. Cherry Burrell's Thermutator provides a scraped-surface heat exchanger that is adaptable for both high- and low-acid foods that are fibrous, viscous and contain particulates. The 'Aseptic Direct Steam Incorporation System' offers all of the advantages of direct heat exchange, that is, short processing time, fast heating and flash cooling. It also uses CB's UHT/Aro-Vac unit, which is capable of heating and flash cooling the product in a fraction of a second, thereby stabilizing product characteristics for delicate formulations such as a complete range of dairy products, including low-lactose or lactose-free milk, ice-cream mixes, milk-shake mixes, creams, cheese sauces, soups, stews and so on.

2.3.5 DASI (TCI-Superior)

TCI-Superior, a Mueller Company, is the sole licensed manufacturer of the patented DASI Free Falling Film system. The principle is based on a film of product falling through an atmosphere of steam, thus ensuring conditions for HTST and preventing a burnt or cooked flavour when the film is milk, and so on. Product is preheated in a plate heat exchanger to 74°C, then enters the infusion chamber of a cone-shaped sterilizer where it falls as a film through an atmosphere of pure steam. Within 0.33 s the product rises to 140°C and is held at that temperature in a holding tube for 3 s. It is then flash cooled to 64°C, homogenized and then chilled to 3.5°C for storage in aseptic tanks prior to packaging. A characteristic feature of the process is that product is never subjected to turbulence or pumping at high temperatures, and never contacts a surface hotter than itself. Products processed by the DASI system can achieve an extended refrigerated shelf-life of up to 90 days. Throughputs range from 2000 up to 8000 l/h. According to Frost and Sullivan, up to the end of 1989 there were eight commercial DASI FFF systems in operation in North America, one in Japan and one in South Korea. The system is used on dairy products by Burger King, McDonald's, Carnation and Syso.

2.3.6 FranRica

FranRica is a broad-based company supplying:

- Surge tanks
- Scraped surface heat exchangers
- Tubular heat exchangers
- Steam injection systems
- Pumping systems
- Valves and connectors
- System instrumentation and control panels

It also participates in aseptic packaging with drum fillers and bag filling systems, for example bag-in-bin, bag-in-drum and bag-in-box systems.

2.3.7 Terlet

Machinefabriek Terlet NV, a Dutch company, is represented in the USA by Manning and Lewis Engineering Company which specializes in heat-transfer systems. The Terlotherm is a vertical scraped-surface heat exchanger for achieving full sterilization or pasteurizing conditions. The heat exchanger has certain unique features, which include:

1. Double heat transfer surface.
2. Close and enhanced scraping to improve heat-transfer efficiency and prevent accumulation of debris on the heat-transfer surface.

The Terlotherm is well suited for handling pumpable viscous products, as well as products containing particulates without them undergoing further damage. There is no advantage with the Terlotherm with products of lower viscosity, in other words where there is no risk of product sticking to surfaces.

2.4 Packaging systems

The most common sterilizing methods and their users are briefly described. The Dole system (by Graham Engineering), used mostly for low-acid foods, relies on superheated steam for sterilizing metal cans. The steam temperature is set at 222°C. Dry hot air is used for sterilizing metal, composite cans and plastic cups suitable for filling high-acid products and the sterilizing environment operates at atmospheric pressure. The Hassia form/fill/seal system uses hot sterile air at 140°C to sterilize plastic cups produced by the thermoforming machine. A combination of hot, sterile air and sterile steam is used by Bosch on their Hypa-S containers, known as 'combi-cans', while hot water and citric acid (pH 4) are used by Crosscheck to sterilize their preformed multi-layer plastic cups. Chlorinated water is used in the Serac system for sterilizing containers (e.g. HDPE, PVC, PET, glass or metal aerosol cans) prior to filling.

FDA regulations in the USA permit hydrogen peroxide concentrations up to 35% by weight to be used in food-processing plant. In practice, solutions of varying strength up to the limit are used, either on their own, or in combination with other sterilants on a variety of packaging systems. As a result of legislation banning residual hydrogen peroxide at levels exceeding 0.1 p.p.m. of the sterilized food product, packaging systems incorporate techniques for removal of this sterilant from the packaging. Hydrogen peroxide is frequently used with UV light and with acetic acid (peracetic acid). Systems using hydrogen peroxide include:

- Cartons, both web-fed and preformed
- Plastic cups, both roll stock and preformed
- Flexible bags and pouches
- Cans and bottles.

Tetra Pak, Combibloc, International Paper, Liquipak and UltraPak all use hydrogen peroxide as a sterilant. Tetra Pak and International Paper use a hot peroxide bath, while Combibloc resort to a peroxide vapour application. Residual sterilant is removed by a variety of methods including squeezing between rollers, forced evaporation or pressurized hot air. Liquipak sterilize the cartons with peroxide solution followed by exposure to UV light and then sterile steam and hot air to remove traces of the sterilant. PurePak's UltraPak system sterilizes the containers at the conversion plant with ethylene oxide prior to shipping. After the carton blanks are formed on the packing machine they are sterilized with a mist of peroxide, which is then removed with hot sterile air.

Hydrogen peroxide solution is used on Bosch's Servac 78AS and the TFA systems where the roll stock passes through a peroxide bath, which is heated and agitated to allow the immersed web to be sterilized on both sides and edges. Excess peroxide is removed with hot sterile knives and the web thermoformed into cups. The system by Genesis (Freshfill) also uses peroxide solution prior to packaging, which is then followed by spraying of the formed containers with peroxide solution and excess removal with hot sterile air. The flexible bag and pouch systems of Bosch's SVA, DuPont's DA-2000 vertical form/fill/seal system and Inpaco's Mark series for forming and filling aseptic bags, also use in-line sterilization with peroxide solution. Liquibox use peroxide and UV light to sterilize the spout of their presterilized bags before filling. In addition, Bosch sterilize cans and bottles in the inverted position with peroxide solution, excess sterilant being washed out with sterile water. In the Remy Type 70 Non-Integrated Aseptic System, bottles (both glass and plastic) are sterilized with a mixture of peroxide and peracetic acid,

Table 2.4 Classification by machine technology

Type	Description	Supplier
Deposit/fill/seal (DFS)	Sterilization of preformed containers (glass and plastic bottles, jars, metal and composite cans, metal aerosols, preformed flexible bags)	Crosscheck, Dole, Remy, Bosch and Serac systems and several bag and pouch packaging systems
Erect/fill/seal (EFS)	Erects a partially formed package, sterilizes and fills	Combibloc
Form/fill/seal (FFS)	Multi-layer/composite web from roll stock formed, filled and sealed	Tetra Pak, International Paper, PurePak, LiquiPak and Hypa-S
	Aseptic bags	Bosch, DuPont and Inpaco.
Thermoform/fill/seal (TFS)	Co-extruded multi-layer plastic sheet fed from roll stock for thermoforming on filling machine; plastic sheet sterilized prior to forming.	Conoffast, Hassia, Servac and Bosch
Blow-mould/fill/seal (BFS)	Package blow-moulded on line, filled and sealed to prevent loss of sterility. Blow-moulded as separate operation and sealed. Sealed sterile bottle then transferred to filler for filling aseptically.	

Table 2.5 Classification by package type

Type	Description	Supplier
Carton systems	Composite cartons in roll stock, precut blanks, or preformed sleeves	Combibloc, Hypa-S (Bosch), International Paper, Liqui-pak (Tetra Pak) TetraBrik Aseptic (Tetra Pak), Ultra-Pak (PurePak)
Plastic cup systems	Co-extruded roll stock with on-line operation for thermoforming cups; preformed cups produced off-line by converter; both types require aseptic lidding units	Bosch, Servac, Conoffast (Continental Can), Cross-check (Mead/Rampart), Freshfill (Genesis), Hassia
Flexible bag and pouch system	Preformed bags and pouches or flexible multi-layer roll stock formed into bags on-line; for bulk packaging outer container used, such as box, bin or drum	Asepak, Bosch, FranRica, Inpaco, LiquiBox Manzini, Scholle, StarAsept (Alfa Laval), DuPont
Can and bottle systems	Designed to handle rigid containers, including metal and composite cans, glass and plastic bottles and jars	Bosch, Dole, Remy (CMB), Serac

known as ‘Oxonia’. Again, excess sterilant is washed out with sterile rinse water. Asepak, Bosch, FranRica, LiquiBox and Manzini supply bags in the presterilized condition using radiation treatment at the convertor’s site. The bags are opened within the sterile environment of the filler and, to facilitate maintenance of sterility, they incorporate a variety of features such as protective membranes, closure fitments, and so on.

At least two companies (Continental Can’s Conoffast and ERCA) operate the ‘neutral aseptic system’ of avoiding use of chemical sterilants by delaminating the web and exposing a sterile surface, under sterile conditions. What remains is then thermoformed into containers. The various aseptic packaging systems may be grouped together according to the type of technology adopted or by the package type. These are summarized in Tables 2.4 and 2.5.

For details of the companies and systems listed herein the reader is referred to the excellent review of the North American market in aseptic food and beverage packaging systems by Frost and Sullivan (1991).

2.5 Future requirements and trends

Much will centre about particulates greater than 15 mm—especially the performance of heat exchangers and their ability to prevent organoleptic changes during heat treatment. Economy drives will improve heat recovery rates, which presently vary between 50% and 90%. Sterilization of particulates between 15 mm and 25 mm presents problems on scraped-surface heat exchangers and requires special handling. The time required to heat, hold and cool product containing large particulates is longer for the liquid phase because of poorer heat transfer.

Consequently, if both the carrier fluid and the large particles are sterilized simultaneously, the liquid phase will be exposed to higher residence times and consequently will be over-cooked. Therefore, the liquid and the particulates require separate treatments using different heating modes. Conventional approaches are singularly lacking in this capability with any high efficiency; this is where the ohmic heating approach developed by APV in England has a worldwide lead.

There is pressure on the machine manufacturers to increase the operating speeds of packaging systems, currently averaging 6000 units/h, without compromising product quality and integrity. Both plastic cups and bricks are likely to show improvements in filling outputs during the 1990s. The main driving force in material development will be centred about replacement of foil barriers with a thermoplastic layer to conform with recyclability requirements as well as improving opening, pouring and reclosing of large containers and improving initial seals, closure integrity and tamper resistance.

On the processing front, developments are expected in new sterilizing agents and their method of use under operating conditions, improving the cost-effectiveness of sterility monitoring (e.g. developing reliable and speedy alternatives to total incubation), enhancing design aimed at facilitating equipment inspection, maintenance and repairs, and also providing heat-resistant, microwavable packaging as part of an in-line microwave-processing approach. Finally and not least, the consumer will continue to dictate a need to develop new food products and formulations.

References

Environmental Policy Consultants (1992) *Packaging Today*, **14,** No. 1.
Frost and Sullivan (1991) *The US Market.*
Leon R. Frotscher Associates (1987) *The European Market for Aseptic Packaging.*
Wheelock, V. (1988) *Food-Borne Disease.* Food Policy Research Unit, University of Bradford.

3 Packaging materials—their properties and criteria of selection

B. I. TURTLE

3.1 Introduction

Packaging materials are an indispensable part of the aseptic processing operation, since they must maintain foods in the stable aseptic condition and also prevent chemical and physical deterioration over the required shelf-life. Hence the main part of this chapter outlines the manufacture and properties of those food packaging materials and containers commonly used in aseptic packaging processes. A comprehensive treatise on the various packaging materials would need a separate volume, but the mainstream types of food packaging are discussed in some detail here and their properties compared. These include open-top tinplate cans, glass jars, flexible packaging, plastics pots and bottles, and packaging in paper and board. The vital importance of maintaining a high standard of seal integrity for aseptic packaging is stressed, and ways of ensuring this discussed.

In a modern urban society, a huge range of goods are required to be available in all parts of that society in perfect condition. To ensure this they must be transported in packaging that has been specifically designed for the purpose. Also, concerning the provision of food, it is no exaggeration to say that without packaging and its attendant marketing and distribution systems modern society could not exist at all.

Nowadays there is no time to go shopping every day for all the foods preserved in the home in cans or bottles or cartons. It would not be acceptable to take containers to the shops for milk or beer or cooking oil, and then wait while the assistants carefully measured out the quantities from bulk containers. Furthermore, the logistics of delivering so much extra perishable food would be daunting—perhaps three or four times as many vans and articulated lorries would be trying to get into our already overcrowded city centres.

One other aspect is the vital question of food poisoning, since early Victorian packaging was dominated by paper bags or newspaper that were inadequate to keep food clean, and street vending was almost totally unhygienic. So these factors contributed to a very high incidence of food poisoning and infant mortality, which would be quite unacceptable today. It is hardly possible to overrate the importance of packaging in keeping wholesome food in good condition and thereby reducing the incidence of food poisoning.

3.2 Pack selection

Before considering the packaging materials themselves, it is instructive to consider the various functions that the retail food package is called on to perform. Although there are more than 100 different aspects to be considered when developing a new package, they may be conveniently grouped under 22 headings as shown in Table 3.1. Generally speaking, 19 of these aspects are technical and three commercial, although some areas will be common to both, such as tamper evidence and easy opening, or cost-effectiveness and shelf-life.

The greatest demands on packaging arise from the requirements of perishable foods such as milk, fruit juice, stewed steak, vegetables in brine, and so on, which have been processed to give long shelf-lives at ambient temperatures. Canned and aseptically packed foods come into this category, and a package for an aseptic food will certainly need to fulfill the above 22 criteria.

3.3. Pack criteria

Table 3.1 separates these criteria into seven areas, namely: (i) product containment; (ii) physical protection; (iii) food safety; (iv) shelf-life, (v) communication of information, (vi) sales appeal; and (vii) cost-effectiveness. A brief comment will be made on each.

Table 3.1 Requirements of a food package

Contain the product
Prevent physical damage
Run smoothly on filling lines
Withstand packaging processes
Handle easily through distribution
Prevent dirt and other contamination
Protect from odours and taints
Resist rodent attack
Stop insect infestation
Be biologically safe, i.e. non-toxic
Be compatible with the foodstuff
Provide sterility
Prevent ingress of microorganisms
Show evidence of tampering
Control moisture loss or gain
Offer a barrier to oxygen
Protect against light
Maintain gas atmospheres, e.g. CO_2/N_2
Communicate product information
Have sales appeal
Open easily
Be cost-effective

3.3.1 Product containment

The need to contain the product in the sense that liquids or powders do not leak out is obvious. However, it also refers to more subtle leakage via wicking through paperboard packs, or loss of coffee or other aromas through plastics containers.

3.3.2 Physical protection

This is again obvious when dealing with fragile foods like eggs or snack foods, but minor impacts on fresh fruits, for example, will release enzymes and lead to browning and softening. Equally important is the adverse effect on sales of damaged packages themselves—even though the product inside is in good condition!

3.3.3 Food safety

Paramount here is the need to ensure that the aseptically packed food retains its sterility, through a package that prevents adventitious contamination by microorganisms. Tamper evidence is also now unfortunately a desirable requirement, in the face of malicious contamination situations. The other aspect of food safety is the avoidance of long-term chronic toxic effects from the food packaging materials themselves. However, there is no difficulty in achieving this provided that the packaging is approved for food use by a governmental agency typified by the European Community, or a research association dealing with toxicity assessment such as the British Industrial Biological Research Association in the UK.

3.3.4 Shelf-life

While for dried foods moisture gain is a major factor in determining shelf-life, atmospheric oxidation, often catalysed by light, is more critical for aseptically packed foods such as milk, fruit juices, or cream soups. Hence a good oxygen and light barrier, as provided by tinplate or aluminium foil, is needed to ensure maximum shelf-life for these products. However, shelf-life will frequently be determined by marketing considerations as less than the technically available maximum. This could occur in situations where, for example, a plastic container preferred for sales purposes could not give the shelf-life of glass or metal packaging. Alternatively, costs might be reduced by deliberately limiting the shelf-life 'use by' date in comparatively expensive chill-storage conditions.

3.3.5 Communication of information

The package will need to tell the purchaser what food is inside it and whose product it is. Apart from this, there is much other information to be passed on to the customer, such as net weight, list of ingredients, batch number, use-by date, nutritional information, and so on.

3.3.6 Sales appeal

The packaging must look attractive and 'catch the eye' of prospective purchasers, and it should also be easy to open and dispense the product from.

3.3.7 Cost-effectiveness

Value for money in packaging is more important than looking for the lowest price. A cheap but dimensionally variable container could cause more down time during production or an increase of 'leakers' in the market place, thereby negating the apparent cost saving. Conversely, spending more to make a pack look more attractive could be justified in increased sales.

3.4 Packaging materials compared

In Table 3.2 the main types of packaging materials are compared, and it is apparent that technically there are many options available in a given packaging situation. In addition, while conformance with all of the 19 technical criteria is not always essential, the three commercial aspects can never be ignored in today's free market. Hence a final decision on which form of packaging to be used will invariably involve marketing and sales considerations.

Table 3.2 Packaging materials compared

Materials	Main technical advantages
Tinplate	Impermeability Container strength Easily heat-processable Withstands vacuum/pressure
Aluminium	Impermeability Lightweight metal easily formed Container axial strength Withstands pressure
Paper and board	Infinite variety of paper types Ease of decoration Infinite variety of cartons and cases Adjunct to all other packaging materials
Rigid plastics	Properties variable Lightweight Choice of container shape In-house manufacture
Flexible plastics	Properties variable Very lightweight Tailor-made sizing
Glass	Impermeability Inert material Product visibility Axial strength Withstands vacuum/pressure

3.4.1 Metal containers

Packaging in metal requires first of all the manufacture of various types of metal sheeting, which is then used in the construction of containers; most of these containers are produced from steels or aluminium and its alloys. The open-top can is the most important metal container in the food industry for both in-can heat-sterilized and aseptically packed foods, and hence will be given most emphasis. However, aseptic packaging may also employ large steel drums, or aluminium kegs and aerosol containers.

3.4.1.1 Tinplate. Tinplate is mild steel, that is, of low carbon content of between 0.1% and 0.5%, which is coated on both sides with tin by electrodeposition. The base steel plate is made by hot and cold rolling cast steel ingots or continuous steel strip down to a thickness of 0.15–0.30 mm. Some steel plate is given a second cold rolling to reduce its thickness by a further 30–40%, when it is known as double-reduced plate.

After cold rolling, the steel strip must be annealed at temperatures in the range of 600–950°C, to relieve stresses and improve ductility. The plate will then receive a light temper rolling to give the required mechanical properties and surface finish. Finally, the sheet steel is led through an electrolytic tinning bath where any desired weight of tin can be accurately deposited on the steel. The weight of tin used on each side of the sheet is normally between 1.4–11.2 g/m^2, and this may be differentially applied to give a higher tin coating on one side.

Mechanical properties of tinplate. These are determined by the composition and method of producing the base mild steel, and the processes employed to manufacture the steelplate. Batch-annealed plate has greater ductility and is more suitable for drawing purposes than continuous plate, which, although stronger, has greater directionality in the rolling direction and a tendency to split under deep drawing. Tinplate is graded according to its 'temper', and there are six grades, where strength increases and ductility decreases from T1 to T6. Table 3.3 shows typical uses for the various temper grades.

Corrosion prevention. The basic properties of the sheet steel and its tin coating provide the main protection against atmospheric and product corrosion. However, this is improved in two ways. Firstly, after the tin coating is applied, a 'passivation' treatment is given using potassium dichromate, which reduces stoving discoloration and product sulphur staining. Secondly, lacquering is widely used externally

Table 3.3 Temper grades

BA	T1–T3	Deep-drawn components—DRD, DWI
BA/CA	T3–T4	General-purpose —can bodies and ends
CA	T5–T6	Stronger general purpose —can bodies and ends
CA	T6	Nitrogenised steel—high-pressure ends

on tinplate to prevent atmospheric corrosion, since this is more economic than increasing the tin coating weight. However, the main use of lacquers is for internal can lacquering, since otherwise aqueous foodstuffs would attack the tinplate to the detriment of both can and product. Most lacquers are synthetic resins, either solvent- or water-based, such as epoxies, acrylics, formaldehydes and vinyls. If optimum performance is to be achieved, care must be taken to oven-cure lacquers to the correct time and temperature specification, and all residual solvents that might taint foods must be removed.

3.4.1.2 Tinfree steel. The same steel substrate used for tinplate may instead be coated with a layer of hydrated chromium oxide and chromium metal via electro-deposition, when it is known as tinfree steel. Due to the very thin layer of chromium surface defects of the steel plate are highlighted, and so it is necessary to use high-quality steel. While lacquering both sides is necessary to prevent corrosion, lacquer adhesion is excellent, and hence the material is commonly used for drawn components such as can ends and caps.

3.4.1.3 Aluminium sheet and foil. Pure aluminium is too soft for can manufacture, and must be alloyed with magnesium and other metals to improve its strength. Sheet is produced in a similar way to tinplate via a process of hot and cold rolling, annealing, and temper rolling. At gauges of 0.30–0.50 mm it is used for the manufacture of food and beverage cans. Aluminium does not suffer the unsightly atmospheric rusting corrosion of steel plate, since it is very stable towards oxidation once a very thin layer of oxide has formed; this layer is not visible to the naked eye. Nevertheless, aluminium will corrode very rapidly in contact with fruit and other acids, salt solutions, or alkaline liquids, and hence is always internally lacquered where it is used for aqueous foodstuffs.

At thicknesses of 0.20 mm and below, the sheet is called foil. It is produced from the thicker sheet by a high-speed cold-rolling process, where each pass through the rolls reduces the thickness by half. At or near 0.20 mm it is suitable for easy opening (EO) ends and bottle caps, and at reduced gauges down to 0.050 mm it is used to make smooth and wrinkle-wall containers.

Foil for flexible packaging. The foil used for flexible packaging has thicknesses from 0.050 mm down to 0.006 mm, and at these thin gauges the sheet must have maximum ductility and softness if cracking is to be minimised. Hence alloying is not necessary, and 'commercially pure' aluminium is used having an aluminium content of 99.3%, with iron and silicon as the main impurities. At these thin gauges it is possible to make significant economies in manufacturing costs by rolling two layers back to back. While the outside surfaces in contact with the mill rolls have a bright appearance, the inside surfaces will be matt; therefore flexible packaging foil always has a bright side and a matt side.

Foil pinholes. At gauges of 0.007–0.015 mm, which are the most important for food packaging, foil will have minute pinholes that may be observed with back illumination by a strong light. While modern manufacturing methods have greatly reduced their incidence, nevertheless a 0.009 mm foil may have as many as 200 pinholes/m^2. However, once the foil gauge has increased to 0.025–0.030 mm, the pinholing is virtually zero.

While the total area of these pinholes is a tiny fraction of the total foil area, nevertheless they are sufficient to allow a significant transmission of water vapour and gases such as oxygen. Hence, where a moisture- and/or oxygen-sensitive food is to be packed, such as instant coffee, thin foils will always be laminated with a plastics film to block the holes and thereby produce an almost impermeable laminate. The plastics films also have the vital function of protecting the foil against creasing damage during handling, which would further increase the size and number of the pinholes.

3.4.1.4 The open-top can

When the Frenchman, Nicholas Appert, published his famous book on the *Art of Preservation*, in 1810, he founded the canning industry, which preserves perishable foodstuffs by heat processing. For the last 150 years the open-top can has been the most important container for the canning industry, and it remains so to this day. It was also the first container to be used for aseptic packaging; milk puddings, ice-cream mix, and other heat-sensitive foods are currently packed in it. Although the can is not so widely used for aseptic packaging as it is for heat-processed foods, it is necessary to consider it briefly for two reasons.

First and most important is its integrity, that is, the can's ability to maintain perishable foods packed inside it in a shelf-stable condition that is free from microbiological contamination. Second is the complete barrier that it affords against all gases and vapours, thus ensuring very long shelf-lives for any foodstuff. While this latter quality is now less important with modern marketing methods, it is still vital for mainline commodity foods throughout the world. The can most certainly remains the standard against which all other alternative containers should be measured.

Open-top containers for aseptic packaging are available in a range of sizes from around 0.25 lb to more than 6 lb. They will usually be cylindrical in style, but flat, elliptical, oblong and pudding-basin types are also available.

Three-piece cans. The traditional can style is the three-piece can, consisting of a cylindrical body and two ends. Sheets of tinplate are cut on slitters to form body blanks, which are then joined at the sides on the bodymaker to form cylinders. After flanging the cylinder, the maker's end is then double-seamed on. Previously, the sideseam join was made via hooked edges on the blanks reinforced by soldering. However, most cans now have electrically welded sideseams giving a far stronger join. There is a marketing advantage also in that the seam area is narrower and cleaner and the appearance of the can is greatly improved.

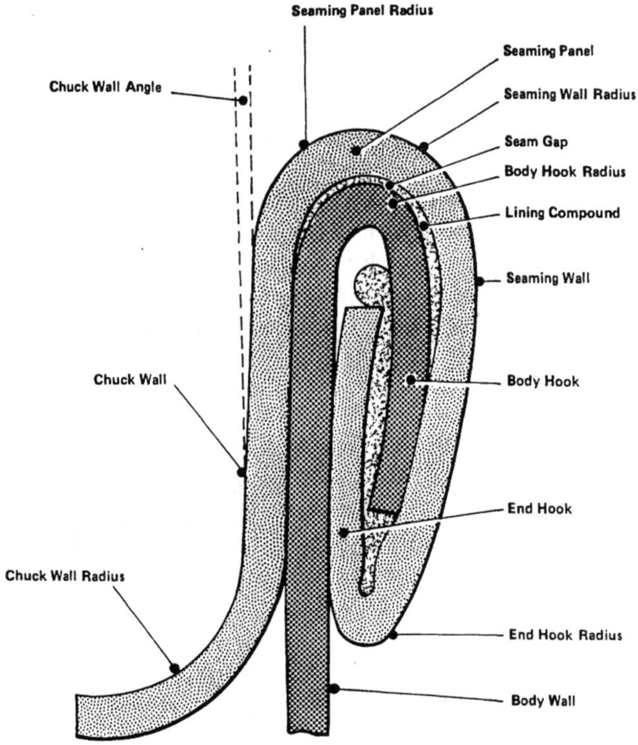

Figure 3.1 Section through double seam (courtesy of CMB Foodcan plc).

Two-piece cans. Cans with an integral formed body and only one seamed end, such as the pudding or sardine-can varieties, have been made for many years by a straightforward stamping method. However, the development of the draw redraw process (DRD), to produce containers with a greater height to diameter ratio, now allows most food cans to be made in the two-piece style. In this process a cupping machine converts coiled lacquered tinplate, or aluminium feedstock, into a circular disc, which is then held in a controlled die gap and drawn into the specified cup shape. The cups then pass to a bodymaker where they are drawn down to the required can diameter, and finally the cylinder end is trimmed and flanged.

From a functional point of view the design of the two-piece can is superior to the three-piece can, since there is only one end to be closed rather than two and consequently less chance of leakage occurring. However, since the can manufacturing lines are around four times as expensive as for welded cans, the two piece will only be economic for long runs on one size.

Double seaming. The double seam that effects the closure of the can is so called because two consecutive operations are required to form it. In the first operation, the curled edge of the end is interlocked with the flange of the can body. In the second, the seam is finished by ironing to the required tightness. Figure 3.1 shows a section of a double seam, together with its nomenclature.

Continuous monitoring of the seams being produced by the double-seaming machines is essential if hermetic seals are to be produced consistently. In particular, the four critical parameters of *seam tightness*, *actual overlap of hooks*, *body hook butting*, and *countersink depth* must be maintained. There must also be an *absence of visual defects*. For aseptic packaging, described elsewhere in this volume under the Dole process (chapter 7), the quantity of rubber sealing compound used in the ends is increased. This is to allow for the fact that the filled can is not heat-treated, and so no swelling of the compound takes place as in the canning process.

The importance of the often repeated phrase that 'a container is only as good as its closure' can hardly be overestimated, since it will apply to all sizes and styles of container, whether made from metal, plastics, glass, or combinations of these materials. The open-top can is no exception to this rule, and its fine reputation as a food container has been built on the excellence of the double-seaming process used to seal an end onto the can body. There is no doubt that the high standard of integrity provided by the can should at least be equalled by any other form of container employed for aseptic packaging.

3.4.2 Rigid plastics containers

The various forms of plastics packaging have such a major role in today's packaging scene that it is difficult to imagine the food industry without them. However, 40 years ago when they were little used in packaging, there was also at the same time an absence of 'convenience' foods such as ready meals, snack foods, prepared desserts, extended-life perishables such as fresh fish or meats in a modified atmosphere, and so on. In the last 40 years the retailing of food has changed from being served by a shop assistant to a self-service business, and competition and the domination by large supermarkets have ensured at the same time that food packaging has become ever more sophisticated. It is true to say, however, that without the continual development of plastics packaging over this same period, the transformation that has occurred in food packaging and packaging systems could not have come about.

While low-density polythene (LDPE) was discovered in the late 1930s, it was not until the early 1950s that film and rigid caps and containers began to be used in packaging. In the mid 1950s high-density polythene (HDPE) and polypropylene (PP) were discovered, and a stream of other polyolefin derivatives have followed, such as linear low-density polythene (LLDPE) and numerous ethylene copolymers. The other major food packaging plastics, such as polyvinylchloride (PVC), high-impact polystyrene (HIPS), and polyethylene terephthalate (PET), also became important from the 1950s onwards. When in the early 1980s techniques were developed to co-extrude different types of rigid plastics, high-barrier pots and trays with low permeability to oxygen became available for the first time. These containers are capable of providing a long shelf-life at ambient temperatures for processed and aseptically packed foodstuffs.

A thorough understanding of plastics and the way containers are manufactured from them is essential if their vital role in modern food packaging is to be appreciated. Apart from rigid plastics containers, the same range of plastics is used for flexible retail packs or shrink-wrapping, as one-piece caps for beverages or for overcaps, as liners for metal caps, and many more applications in conjunction with other packaging materials. Hence the properties of plastics are dealt with in some detail in the following pages.

3.4.2.1 Polymerisation. When a reactive monomer such as ethylene, propylene or styrene is subjected to heat and pressure in the presence of a catalyst, the monomers link together to form polymers. 'Poly' means many, and 'mer' comes from the Greek 'meros' meaning a part, hence the name polymer for the long-chain molecules formed by the polymerisation process. The basic average chain length, and the configuration of the chains, will vary dependent on the monomer and the polymerisation process. The chains may also have branches characteristic of the material so that, while HDPE has short side-chains and is a comparatively stiff plastic, LDPE is more flexible with long branches having further subdivisions. Cross-linking to give a stiffer plastic may be engineered in many cases during polymerisation, or it may be produced afterwards by treatment with ionising gamma radiation.

Another important attribute of plastics is that, within a particular chemical type, for example PVC or LDPE, the polymerisation process may be adjusted to vary the properties of the material produced. Both the average chain length and the chain distribution can be altered, thus producing polymers of varying degrees of processability to suit the different methods of making containers. In addition, copolymerisation of two or more monomers will produce different properties. Polypropylene copolymerised with ethylene, for example, is less brittle and easier to seal than straight polypropylene.

3.4.2.2 General properties. The properties of individual plastics are mentioned later, but there are many characteristics common to all types that are conveniently described as follows. Compared with metal and glass they are lightweight, with densities around 1 g/cm^3, compared with iron 7.9 g/cm^3, aluminium 2.7 g/cm^3, and glass around 3 g/cm^3. They are, in general, flexible with good impact strength and rarely shatter destructively, although at deep-freeze temperatures of –20° to –30°C the impact strength decreases markedly for some plastics, such as polyvinylchloride or polypropylene. Nevertheless, modifications to the material via copolymerisation or the addition of rubber particles will solve any strength problems that an individual plastic may exhibit.

Plastics are free from corrosion problems and chemically inert, but they have drawbacks such as dust pick-up due to static electricity, and also the poor size impression they give relative to glass bottles and jars. However, one of the main strengths of plastics is the ability to vary shape and size, which gives the designer much scope to vary the form of the container. This attribute, together with technical versatility, is largely responsible for their widespread use in food packaging.

Plastics transparency. The clarity of plastics depends on the presence or absence of chemical or physical groupings capable of absorbing visible light. General-purpose polystyrene is amorphous in structure and crystal clear, while LDPE is largely amorphous but with some crystalline regions consisting of closely aligned spherulites, which scatter visible light and prevent glass-like clarity; however, by rapid quenching when the polythene film is produced, smaller spherulites are formed and clarity is much improved.

Permeability. Plastics all have measurable permeabilities to gases and moisture and other vapours, and this is generally a disadvantage in reducing the potential shelf-life of a food, or allowing tainting by external vapours. Nevertheless, in some instances, this permeability can be made use of, examples being the controlled drying out of cakes to prevent mould growth, or the modified atmosphere packaging of fruit and vegetables. While vegetable oils can soften HIPS containers sufficiently to cause cracking under top-load, or permeate through LDPE, nevertheless there are no problems with regard to attack on plastics by any foods, provided that the correct choice of plastic is made in the first place. Since they absorb mobile food components such as flavours or fats, and tend to be easily abraded, they are not particularly suitable as re-useable containers.

Thermoplasticity. Virtually all the plastics used for food packaging are thermoplastic in nature, that is, they may be formed or reformed by the application of heat in a controlled way using metal moulds or heating bars. The temperatures required to do this will vary with the type of plastic and the particular process, but tend to be in the range of 120–220°C. The softening point of polyvinylchloride is around 70–80°C, low-density polythene 85–95°C, high-density polythene 120–130°C, and polypropylene 140–150°C. These temperatures will not only influence the processing conditions needed to form them, but will also determine the in-use temperature of the finished plastics containers. Both high-density polythene and polypropylene are suitable not only for pasteurisation around 100°C, but also for full sterilisation at 115–130°C, although only polypropylene can be used at the highest temperatures.

The ability of plastics to soften under heat is extremely valuable for closing flexible pouches, or enabling containers such as plastic pots or bottles to have heat-sealed diaphragms. When similar plastics surfaces are brought together under heat and pressure they fuse to give a hermetic seal, which is stable when cooled. The subject of heat sealing is dealt with separately in this chapter, as it is so vital to the integrity of plastics containers and cartons for aseptic packaging.

3.4.2.3 Plastics for foods. In general, plastics are not subject to attack by microorganisms and do not support their growth, and this is particularly beneficial with regard to their use for aseptic packaging where there is no post-filling sterilisation procedure.

Low-density polythene. LDPE is the most widely used packaging plastic, as it is suitable for pots, bottles, caps, flexible films and bags, and as an adhesive on cartons or in flexible laminates. It is, in fact, a family of materials, ranging in density from 0.915–0.930 g/cm^3, that provides variation around the basic characteristics of polyethylene.

The moisture barrier it affords is good but its barrier to gases, oils and aromas is poor. Copolymerisation with Surlyn ionomers gives better resistance to fats, however, and also increased puncture resistance. Another copolymerisation, with vinyl acetate (EVA copolymers) increases the impact strength and flexibility, and reduces the heat-sealing temperature. Linear low-density polythene (LLDPE) has better tensile strength and elongation than LDPE, and a greater resistance to stress cracking. EVA and LLDPE are used more in the film form than as three-dimensional containers.

High-density polythene. HDPE has a more linear structure than low-density polythene, and is stiffer. It is an excellent barrier to moisture, with better gas-barrier properties than LDPE. It may be subject to environmental stress-cracking in contact with the essential oils found in flavourings, but this can be avoided by the choice of a crack-resistant grade of polymer. The clarity of the material is very poor, and at best may be described as see-through.

Polypropylene. Polypropylene (PP) is a harder material than the polythenes, and its low density of 0.9 g/cm^3 gives better yields and improved economics. It has excellent resistance to flexural fatigue, and loading stresses that could cause 'creep' distortion of the material; also it does not stress-crack. Copolymerised with polythene it has good impact strength and seals easily.

While moisture- and gas-barrier properties are similar to HDPE, they may be improved by a process known as orientation. After polypropylene film is made it is stretched while still hot in the machine and transverse directions to align the polymer molecules closer together. In addition, the toughness of the film is increased and clarity improves from good to 'glass-clear'. Orientation of containers is also carried out, using the stretch-blow process described later. While orientation was pioneered with polypropylene (OPP), it is now used to improve other plastics such as polystyrene, PVC and the nylons.

Polystyrene. General purpose polystyrene is crystal clear, but due to its brittleness and ease of fracture it is mostly used in food packaging when modified with butadiene rubbers. It is then known as high impact polystyrene (HIPS), and because this material has poor clarity it is invariably used in the pigmented form. A terpolymer, acrylonitrile butadiene styrene (ABS), having better resistance to stress-cracking, is used for tubs for soft margarines and other fatty spreads. Since the barrier properties of styrene are poor, multipolymers such as Barex or Lopac have been developed having a good barrier to gases combined with clarity and toughness.

Polyvinyl chloride. Polyvinyl chloride (PVC) has good clarity and offers a reasonable barrier to gases and aromas, but it is only a moderate barrier to moisture. Containers usually contain modifiers such as the acrylics to improve the impact strength. PVC or its copolymers are also plasticised with esters such as dioctyl phthalate, to soften the rigid material and produce stretch or cling films. A similar chlorinated polymer, polyvinylidene chloride (PVDC), cannot be made into containers but is valuable as a coating material to improve the gas and moisture barriers of other plastics.

Polyester. Polyester (PET) has better barrier properties than PVC and improved light transmission, particularly in the oriented state where it exhibits glass-like clarity. It is a tough material in both container and film forms. If amorphous clear PET is heated above 150°C under pressure, the PET changes to the opaque crystalline form (CPET), and the containers produced may be heated in domestic ovens at temperatures up to 260°C without distortion. Polyester bottles made by the stretch/blow/moulding method were first introduced in the USA in 1978 for carbonated drinks, and the worldwide market is currently measured in billions of bottles. Containers suitable for aseptic packaging are now also made by this process.

Polycarbonate. While its barrier properties are poor for food packaging applications, polycarbonate has a high impact strength and glass-like clarity. Heat resistance is excellent and it may be steam sterilised at up to 130°C. It is now available in film form, in addition to three-dimensional containers.

Polyvinyl alcohol. Polyvinyl alcohol (PVA) provides an excellent barrier to gases, vapours and aromas second only to metallised laminates. However, the pure material is dissolved by water, so unless a soluble sachet is required to perhaps add flavourings or colours to a food mix, it will be employed as a copolymer of vinyl alcohol and ethylene. At ethylene contents in the copolymer of around 30%, a balance is struck between the retention of a very good gas and aroma barrier while gaining the necessary resistance to moisture for the packaging of aqueous foods. A widely used copolymer has the trade name EVAL, and is of importance in the manufacture of high-barrier containers for aseptic packaging discussed later in this chapter.

Nylons. There are several different nylons used in food packaging but because of their high cost they are used in film form rather than as containers. Nylons are produced via condensation rather than free-radical polymerisation. The first nylon was nylon 6.6, prepared by the condensation of adipic acid and hexamethylene diamine, while nylon 6 is an internal condensation of caprolactam. All the nylons have good gas-barrier properties with excellent clarity, and their high temperature resistance and toughness make them particularly suitable for boil-in-bag meals. A more recently developed nylon, MXD6, is said to have a gas barrier at least ten times better than conventional nylons plus reduced sensitivity to moisture.

3.4.2.4 Gas and vapour transmission of plastics. All the thermoplastics used for food packaging are permeable to a measurable extent to moisture vapour, oxygen, and all other vapours and aromas. Dry foods are particularly vulnerable to moisture gain or loss, such as in the softening of biscuits or the shrivelling of fruits and vegetables. Aqueous products, however, such as aseptically packed fruit juice or full-cream milk, will have their shelf-lives severely limited due to oxidative rather than moisture changes if the packaging does not limit the ingress of oxygen.

Oxygen and moisture transmission rates. Table 3.4 lists the oxygen and moisture permeabilities of the food-packaging plastics; it should be noted that, while perfectly adequate as a guide to plastics selection, these are typical values rather than exact parameters. There are so many variations possible within any particular family of plastics materials that precise figures can only be quoted for an individual supplier's resin—or better still for the container that is being considered for use. Where there is also a need for some specific property, such as heat resistance or clarity, then lamination of individual plastics will enable the required transmission rates to be obtained at the same time.

Other gases. Nitrogen and carbon dioxide are frequently used to extend the shelf-life of food. Nitrogen may replace headspace atmospheric oxygen to limit flavour and colour changes in aseptically packed foods. Carbon dioxide is widely

Table 3.4 Oxygen and moisture transmission rates with 25 μm material thickness

Material	OTR ($cm^3/m^2/24h/atm$ at 25°C and 50% RH)	MVTR ($g/m^2/24h$ at 25°C and 75% RH)
PVDC	2	1
EVOH	0.5–10*	40
MXXT/A Cello	6	1.2
Acrylics	15	25
PVDC-PVC	15	9
Nylon 6	50–150*	40
Polyester	80	8
PVC	200	20
HD polythene	1400	2
OPP	1500	2
ABS	1600	35
Polycarbonate	2000	40
HIPS	4500	30
LD polythene	8000	5

* Humidity dependent

used as a bacteriostat for perishables such as fresh fish and meat, and also to extend the shelf-life of cakes and pastries packed under 'clean' if not completely aseptic conditions. Nitrogen is less permeable and carbon dioxide more permeable than oxygen for the packaging plastics, and while different plastics types show variations, this ratio approximates to an N_2:O_2:CO_2 ratio of 1:4:14.

3.4.2.5 Container manufacture. Rigid plastics bottles, pots, jars and trays are available in an almost unlimited range of sizes and shapes with capacities from 5 ml to 20 l. They may be closed in a variety of different ways, such as by heat-sealing on a diaphragm, double-seaming a metal end, by the use of a metal or plastic screw-closure cap, or by sealing the container walls to each other. This latter approach has been used when filling milk into plastics bottles, the top part of the neck being fused on itself; both pasteurised and aseptic UHT milk has been packed in this way.

To accommodate this wide range of container styles there have inevitably developed numerous ways of making the containers, and this process is a continuing one as consumers and food manufacturers demand ever higher standards of quality. Nevertheless, the mainstream of manufacturing is still based on three methods and variations around those methods. They are: (i) injection moulding; (ii) blow moulding; and (iii) thermoforming, and the salient points of these processes are described briefly.

Injection moulding. In a simple screw injection moulding machine, the hopper of the machine contains cold granulated or powdered polymer, blended with additives such as colourants, antistatic agents, impact modifiers, flow improvers, and so on. The rotating screw draws material into the barrel of the machine, where a combination of friction and electrical heating softens or 'plasticises' the mix. The screw, which has retracted during this operation, now moves forward to inject the plastic into the shaped mould. After a cooling period the mould opens, the plastic container is ejected, and the process begins again.

Injection moulding is capable of producing pots and trays to a very high dimensional standard. However, to achieve this, the moulds have to be engineered to very narrow tolerances from high-quality steels and they are also required to stand high pressures on the injection cycle. Consequently, multi-cavity moulds are expensive and require long runs to be economic.

Thermoforming. A sheet of plastics material is first softened by radiant heat from infra-red heating elements, heating from both sides for the faster machines or thick sheets. The sheet becomes sufficiently soft to be formable, although below its melting point so that it may be handled by the machinery. At the forming station, the sheet is clamped and then drawn into the mould by vacuum or air pressure, or often a combination of the two. For deeper-drawn containers a metal plug will be used to help push the sheet into the mould cavity. Figure 3.2 illustrates the thermoforming principle.

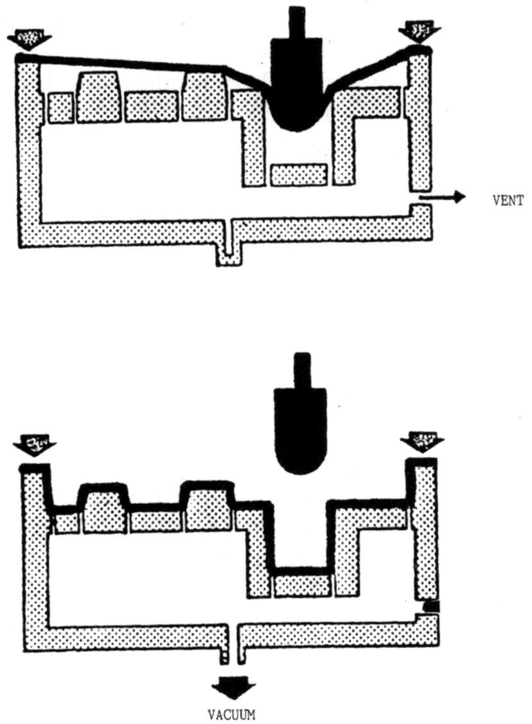

Figure 3.2 Plug-assisted thermoforming.

There are some limitations to the thermoforming process. One major constraint is that the pots and trays produced cannot have more than a minimal degree of reverse wall taper, otherwise they are difficult to remove from the mould without distortion. There is also more inherent variation in container wall thickness than for injection moulding, although improvements in material and forming technologies have greatly reduced this. Thermoforming has now become the most widely used method for producing containers for clean and aseptically packed foods, such as soft yellow fats, yoghurts and desserts.

Blow-moulding. There are three methods of blow-moulding bottles and wide-mouth plastics containers: (i) extrusion blow-moulding; (ii) injection blow-moulding; and (iii) stretch blow-moulding. The most widely used technique is the extrusion route and, as for injection moulding, a rotating screw in the electrically heated barrel of the extruder softens the plastics material. The plastic is then extruded through a tubular die of a size to suit the container being blown, and the tube clamped between the two halves of a metal mould. This tube is then blown out to the shape of the mould by air pressure, and after a period of cooling the container is ejected from the mould. The thickness of the extruded tube is varied to give precise control of container wall thickness and so put material only where it is needed, thus achieving minimum weight and thereby minimum cost.

Injection blow-moulding entails the production of a thick-walled parison that has a fully formed neck ring. The hot parison is then transferred to a blow mould to produce the final bottle dimensions. This method gives a fully finished bottle or container together with precise control of the neck ring thread and other dimensions, but the machines are more expensive than for extrusion blow-moulding.

Stretch blow-moulding. This is an important variant on the other two blow-moulding processes, in which a preformed parison is stretched by a telescopic metal rod during the container blowing stage. The containers are blown at temperatures well below the melting points of the plastics; this ensures that a considerable degree of orientation is produced. The advantages to be gained from orientation are higher tensile and impact strength, reduced creep, increased clarity, and improved barrier to gases and vapours. The polythenes, polypropylene, polystyrene and other types of plastics can all benefit from orientation.

3.4.3 High-barrier plastics containers

Nearly all aseptically packed foods, such as milk, fruit juices, prepared desserts or soups, will need to be protected against atmospheric oxidation, and the light that catalyses this reaction. However, examination of the various food-packaging plastics indicates that one material on its own is unlikely to provide a sufficient oxygen barrier together with good moisture resistance and strength, at an economic price.

Polymer blends such as nylon with high-density polythene have been investigated, and a blend of PET with MXD6 nylon plus a cobalt catalyst has been found to have unique oxygen-scavenging properties (the Oxbar System from CMB plc). Nevertheless, for general use, the best way to combine a high barrier towards oxygen with other desired properties is by coating the formed containers or making them from co-extruded materials. Coating with PVDC is widely used on PET bottles for beverages, where it reduces gas permeability by at least 50%; also a new technique, which entails the high-vacuum deposition of silicon dioxide on to finished containers, is said to reduce permeability still further (QLF (quartz-like film) from Eastpac plc). There is much current development to improve oxygen-barrier properties since this is an important growth area for plastics packaging in the next decade.

3.4.3.1 The co-extrusion process. This technique involves the production of either a multilayer tube or sheet, to be blow-moulded into a bottle or thermoformed into a tray respectively. Several extruders are used to feed different plastics into a central die where they are combined together. There are several ways of doing this, but one of the most successful is the feedblock system (see Figure 3.3). In this method the different layers flow on top of one another, and are combined within the block before emerging as a single sheet. Most of the laminates incorporate

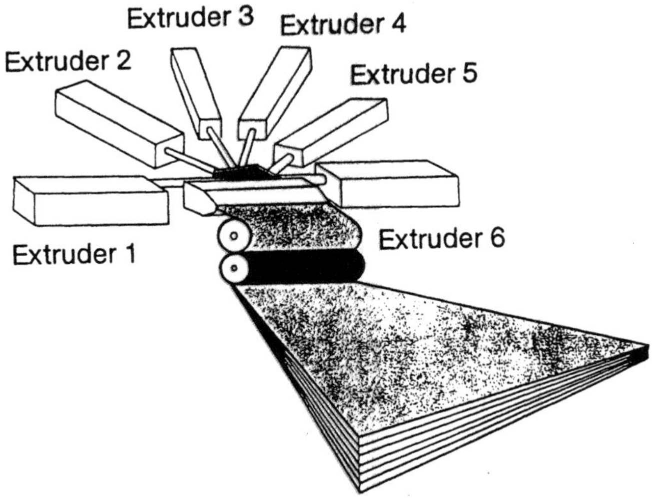

Figure 3.3 The feedblock system (courtesy of Cobelplast SA).

adhesive or "tie" layers to ensure good ply adhesion. A laminate sheet, having an oxygen permeability as low as 0.5g/m^2/24h/atm at 25°C, might have 6 layers consisting of, from the outside, PP/Regrind/Tie/EVOH/Tie/PP, though there are many variations around this typical structure.

3.4.3.2 In-house manufacturing. Before leaving the subject of plastics containers mention should be made of in-house or in-plant manufacture. Of all the container materials, plastics are unique in that the use of comparatively low-cost manufacturing equipment is feasible for all but the smallest of food packers. In addition, plastics pot suppliers often charge tool and origination costs anyway. Hence the economics of own manufacture are often very favourable.

There are many factors, apart from the basic machinery costs, to be considered before starting own manufacture. Extra space is required, away from food preparation, for manufacture and storage of raw materials, and also for ancillary equipment such as a printer or a tray packer. Other aspects requiring careful analysis are continuity of supply in the event of a breakdown, and the question of increased output for peak periods or an unexpected runaway success for the product.

However, there is now a widespread employment of in-house manufacture using thermoforming machinery, and the main reasons for this are as follows:

1. Thermoformed containers have an intrinsically low cost;
2. Thermoforming machines are available for £0.5–£1m, depending on speed and mode of aseptic operation;
3. The food manufacturer is not required to monitor and process raw plastics materials, since sheet to form the pots is available in monolayer or co-extruded forms from many suppliers;

4. The equipment is built on the 'form/fill/seal' principle, which integrates the manufacturing operation with the subsequent filling of the foodstuff and sealing it inside the container; and because of
5. The increasing importance of the aseptic packaging technique, for which the horizontal form/fill/seal machine is ideally suited.

3.4.4 *The special needs of plastics in aseptic packaging*

The numerous requirements of a successful food package were outlined in Table 3.1, and it was noted that packaging for aseptics was particularly demanding of the pack. Some of these demanding aspects, such as long shelf-life, high seal integrity, or consumer appeal, are shared with other areas of packaging and are discussed later in this chapter. However, because plastics materials are so important to aseptic packaging, it is useful to consider the main special properties demanded of plastics by the aseptic process itself. They are as follows:

1. Chemical resistance and wettability;
2. Thermal stability;
3. Low levels of contaminating microorganisms; and
4. Resistance to ionising radiation.

3.4.4.1 Chemical resistance and wettability. The most widely used method of sterilising the surface of plastics containers involves hydrogen peroxide in aqueous solutions of 30–35%, followed by a blast of hot air at 150–250°C. While no major chemical attack on the plastics materials is possible, given the short contact time, surface oxidation could conceivably produce molecular fragments causing tainting of the food if unsuitable materials were used, for example those containing excessive low-molecular-weight polymer. The same comments apply similarly when the hydrogen peroxide is used in conjunction with peracetic acid.

Another important point is that the peroxides must thoroughly wet the surfaces of the containers if they are to contact the microorganisms to be destroyed. This may require the use of surface treatments such as corona discharge on some polyolefin materials. However, plastics such as PVC or polystyrene, which are polar in their chemical nature, will not need modifying in any way. Nevertheless, the level of wettability must not be such as to make removal of the residual hydrogen peroxide difficult for two reasons: firstly, because food safety laws prohibit residual levels above 0.5 p.p.m. in the food, and secondly because excess residual hydrogen peroxide could give food tainting.

3.4.4.2 Thermal stability. Unlike metal and glass, which are unaffected by the temperatures used in heat sterilisation, plastics containers all soften to a greater or lesser extent depending on the temperatures being employed. While PVC and HIPS with softening points 70–80°C and 80–100°C respectively would not withstand thermal processing, they are well suited to the short exposure to high temperature required of the hydrogen peroxide sterilisation technique. This is one of the main reasons why hydrogen peroxide is so widely used at present, in combination with either hot-air or infra-red heating.

However, over the past decade, grades of polypropylene that are far easier to thermoform than hitherto have been developed. This means that thin-wall polypropylene containers are now much more competitive with HIPS for chill-cabinet foods such as yoghurt and cream. With their high softening points around 140–150°C they can be used with aseptic installations employing superheated steam at 130–140°C as the sterilising agent. The attractions of steam sterilisation are the reduced cost, since steam is cheaper than hydrogen peroxide, and the fact that sophisticated monitoring techniques are not required for residual sterilant levels. However, care must be taken to avoid excessive steam condensation into the product. Hassia and other machinery manufacturers now have steam sterilisation form/fill/seal units that can use polystyrene materials; they achieve this by holding the containers in a water-cooled mould while the steam flushing occurs, thus preventing distortion.

3.4.4.3 Low levels of contaminating microorganisms. Whether the method used to sterilise the internal surfaces of containers for aseptic foods uses hydrogen peroxide or superheated steam, good manufacturing practice will require that 'overkill' is not employed. This would only increase the risk of attack on the plastics and/or tainting of the food. However, because the number of bacterial spores surviving a particular sterilising procedure depends on the initial population, any optimum process that is established relies on a predetermined consistently low level of initial contamination. An 'acceptable' value for this 'low level' is often quoted as a maximum of 10 organisms per fluid ounce, but the standard is invariably higher than this.

As a result of the high temperatures employed, the surface of a plastics container that has just been fabricated is sterile, and so form/fill/seal operations should have no contamination problems. Where preformed containers are used, however, it is necessary: (i) for the manufacturing plant to have good hygiene practices; and (ii) for the containers to be supplied to customers within sealed polythene bags inside the outer cases.

3.4.4.4 Resistance to ionising radiation. Ionising radiation may be defined as 'any type of radiation that is sufficiently energetic to induce the formation of positive and negative ions in the treated substrate'. When suitable doses of radiation are applied to microorganisms and insects, their DNA molecules are inactivated, and consequently death occurs. The radiation sources commonly used are: (i) radioactive cobalt-60, which emits gamma rays; and (ii) an electron beam generator, which produces a stream of high-energy electrons. Gamma rays are more penetrating than electron beams, but the received dose lethal effect is the same for either type of radiation, and they are both used to sterilise packaging materials.

Uses of irradiation. Irradiation is widely employed for the sterilisation of packaged surgical instruments and other hospital medical devices, and also food for patients on sterile diet regimes. It is also used to sterilise large flexible bulk containers for liquids made from polythene or polyolefin laminates, and for on-line sterilisation in form/fill/seal operations. However, apart from complete sterilisation, irradiation is also a valuable means of reducing the initial population of microorganisms on packaging materials to ensure that the aseptic process produces consistently sterile products. Applications of this technique include polythene-coated carton blanks for milk and juices, polystyrene pots for single-serve milk or cream portions, and polypropylene pots for dairy products.

Gamma irradiation is used to produce cross-linking in polythene and other materials, thereby improving toughness and shrinkfilm orientation stress-release properties. However, some breakdown will occur both in polythene and other plastics when they are irradiated. PVC, particularly when plasticised, may produce hydrogen chloride or carbon dioxide, and all the thermoplastics will degrade to give lower-molecular-weight fractions.

Biological safety. The possible harmful effects of such degradation products migrating into foodstuffs have been investigated by numerous authorities, such as the US FDA and the European Community's Joint Expert Committees on the Wholesomeness of Irradiated Foods (JECFI). They concluded that at the low level of irradiation permitted, 10 kGy or less, there were no harmful effects. Tainting by these breakdown products is a possibility that is being investigated. However, with regard to the pretreatment of aseptic packaging materials, the dose level of 3–7 kGy is lower than that required when sterilising the foods themselves, and there are no off-flavour problems with the many systems currently operating throughout the world.

3.4.4.5 Plastics and aseptic packaging. The special needs of aseptic packaging have been considered with respect to plastics because they are the most widely used materials for this purpose, either in monolayer form or as rigid or flexible laminate containers. And they are certain to be employed even more extensively as the importance of the aseptic techniques themselves increases. Given the extensive choice of plastics currently available, with new types continually being developed, it is prudent to understand the problems that could arise if the testing of new containers and/or practices were not researched thoroughly. However, at the present time, there is no operation involving irradiation for aseptic packaging for which suitable plastics containers are not available.

3.4.5 Glass containers

Glass is one of the earliest packaging forms, although its use for packaging did not truly start until the Venetian glassblowers in the 16th century began making wine bottles on a large scale. In 1903 the first fully automatic glass-making machine was developed, and in the 1920s and 1930s metal capping systems such as the Omnia and Whitecap were introduced for pasteurised and sterilised foods, so that glass containers could be used for all types of foods and beverages.

Owing to its long history allied to its uniqueness as a packaging material, glass has always been well received by the consumer. Moreover, the competition from plastics and liquid-holding cartons has given an impetus to developments in the fields of strength and lightweighting, which will assure glass a successful future.

3.4.5.1 Composition of glass. The raw materials of glass consist of sand, soda ash, limestone, magnesia, and alumina, plus colourants. Depending on the particular production requirements and the availability of scrap glass, the mix may also contain between 15% and 30% of cullet. This not only saves material costs but also considerably reduces the energy required to melt the mix, thereby making a valuable saving in fuel costs.

3.4.5.2 Container manufacture. The production of glass wide-mouth containers is now a highly automated and microprocessor-controlled operation, which operates on a continuous 24-h basis. There are five production stages, consisting of: (i) glass melting; (ii) feeding and gob forming; (iii) container forming; (iv) annealing and surface treatment; and (v) inspection.

Container forming. After thorough mixing, the raw materials are melted and a 'gob' of molten glass at around 1100°C is transferred to the bottle-forming machines. There are two basic methods of producing glass containers—'blow-and-blow' and 'press-and-blow'— and both have two stages. Glass wide-mouth containers for aseptic packaging are made by the latter method where the gob is pressed into a parison mould with a metal plunger followed by air pressure. The parison is then transferred to the finishing mould to be blown into its final shape.

Annealing and surface treatment. When containers leave the forming machine they must be annealed to reduce stress patterns by passing them through a tunnel known as a 'lehr'. Here the temperature is first raised to near the softening point of the glass around 600°, and then slowly cooled at a controlled rate.

Glass containers have two surface treatments, known as hot and cold end coatings because they are carried out at the start and end of the lehr tunnel. The hot treatment is usually with vapourised tin or titanium compounds, and this greatly increases the shock resistance of the glass. The cold one consists of spraying on slip agents, such as polyethylene glycol derivatives, which reduce the surface coefficient of friction so that containers run more smoothly on production lines.

3.4.5.3 Glass container characteristics. The most important advantages of glass are undoubtedly its transparency, inertness and its barrier properties. From a packaging point of view, the clarity of glass means that a product in a bottle or jar may be seen as easily as if it were not enclosed, which can be a valuable marketing tool. The chemical inertness of glass means that there is no need for internal coatings to prevent interaction between the foods and the glass surface. With regard to barrier, glass is completely impervious to all gases and vapours, and pigmentation can give any degree of light barrier necessary.

Strength. While the axial strength of glass is excellent, its main drawback is brittleness and a consequent poor resistance to impact shock. This brittleness is reduced by good design, which streamlines the shape so far as possible and puts material where it is most needed.

Lightweighting. The average weight of glass containers has been reduced considerably, together with improved impact strength, by attention to the above design factors, coupled with improvements to the press-and-blow technique. A good example of lightweighting is the 1-pint milk bottle reducing in weight over the last 20 years from around 400 g to the 'Redfearn Pintie' at 225 g.

3.4.5.4 Closures for glass. There are a great many different types of metal and plastics closures suitable for glass containers, and most of them will also be used for plastics containers. To ensure good seals, the neck finishes of the jars will conform to a specification demanded by the nature of the closure. Nevertheless, wherever possible, closures will be designed to seal positively over a fairly wide dimensional range, for example, the PT Whitecap. Tamper-evident closures with integral tear bands or added shrink sleeves are now very widely used.

3.4.6 Flexible packaging materials

Flexible packaging consists of lightweight materials in flat form that are made into bags, pouches or sachets, and tend to take their shape from the product inside them rather than being manufactured as three-dimensional containers. However, the heavier flexibles, such as thermoformed trays or paper cartons, are semi-rigid and tend to blur this distinction between flexible packaging and rigid containers. The materials of flexible packaging are papers, Cellophane, aluminium foil and plastics films such as polythene, polypropylene, polyester, and so on.

Paper may be considered as the basic material of flexible packaging, but the earliest papers based on cotton or esparto grass were expensive, and only used for writing purposes. However, in the mid 19th century paper-making machines were developed using wood pulp as the raw material, and they reduced costs to the extent that the use of paper for wrapping became economical. The first 'modern' packaging material was Cellophane in 1912, foil laminating followed in the late 1930s, and then in the early 1950s the introduction of polythene-coated papers accelerated the expansion in the use of flexible packaging to its current important level. The properties of aluminium foil and the plastics materials have been described earlier in this chapter, so only papers and Cellophane are dealt with here.

3.4.6.1 Paper. Although there are a great many varieties of both paper and board, they are all made from cellulose obtained from wood pulps. Papers have a substance not greater than about 200 gsm (g/m^2), and above this weight are termed 'boards'.

Manufacture of paper. The logs are first pulped either mechanically or chemically to produce a slurry, to which are added bleaching agents, colourings and other additives, depending on the type of paper or board it is intended to produce. Packaging papers will be made via the chemical process, because mechanical pulping produces a poor-quality paper with a high percentage of non-cellulose impurities. The diluted pulp is then fed to the Foudrinier machine, which consists of a horizontally travelling wire mesh through which the water drains to produce a web. As paper thickness increases it becomes more difficult for water to drain away, and hence for all but the thinnest boards a cylinder method is used. This technique consists of forming separate thin webs on wire mesh cylinders, which are then combined in pressure rolls to give the board.

At the end of the wire mesh, the web still contains about 85% water, which is then reduced to about 3–4% by squeezing in pressure rolls followed by steam-heated rolls. Sizing or coating treatments are usually carried out during the drying stage, followed by calendering with various rolls to give the surface finish required. Finally the web is moisturised to around 7%, since a 3–4% level would give an unsuitably brittle paper.

The types of paper and board used in packaging are as follows: (i) kraft; (ii) greaseproofs; (iii) folding boxboard; (iv) solid fibreboard; and (v) corrugated fibreboard.

Kraft. The strong brown papers produced by the sulphate or Kraft process are ideal for bags or multiwall paper sacks, and are also extensively used as facing on corrugated fibreboard. While not as strong as Kraft, bleached Krafts have better colour and will be used where this is important.

Greaseproofs. These are translucent porous papers, and vegetable parchment is a high-quality greaseproof treated with sulphuric acid to improve water resistance. Glassine is a transparent paper with better barrier to water vapour and gases.

Folding boxboard. This is a type of board used to manufacture food and other cartons via the cutting and creasing method. This technique can handle boards of a substance from 200 gsm to around 450 gsm, in other words approximating to 300 μm to 800 μm in caliper. Most of the boards will be coated or lined types, and the most important for food is white lined chipboard. Chipboard consists of waste paper and board mixed with virgin pulp and resins and is unsuitable for direct food contact. However, if during manufacture a bleached high-quality pulp is applied to its surface it is then suitable for food.

Solid fibreboard. This is constructed from up to five plies of chip or other board followed by lining with Kraft paper. It is stronger than corrugated board but tends to be more expensive.

Corrugated fibreboard. This board gets its name from the fluted or corrugated structure, which gives strength and shock resistance with lighter weight. The basic type comprises a single fluted sheet with board on one side and a Kraft liner on the other. However, double- and triple-wall constructions are also used, and corrugated cases can be produced in a wide variety of strengths. Specifications relating the size of the flutings, grammage of boards, and so on, to strength are published by major bodies such as the BSI and ISO.

3.4.6.2 Cellophane. Cellophane is a trade name for regenerated cellulose film, which is made from selected trees supplied to the film manufacturer in the form of sheets of wood pulp. In the Viscose chemical process the fibrous wood pulp is 'regenerated' into a non-fibrous transparent form of cellulose. Plasticisers such as glycerine or glycols are incorporated to impart the necessary degree of flexibility, and the film may be dyed or pigmented during manufacture.

Cellophane has a good barrier to gases when dry but is very sensitive to moisture, and so it is mostly used with coatings such as nitrocellulose or polyvinylidene chloride (MXXT film) to impart a moisture barrier. Grease resistance is good and it is easy to handle on packaging machinery.

3.4.6.3 Plastics films. Virtually the same range of thermoplastics described under the containers heading are used for flexibles. However, unlike rigid plastics where materials such as PP, HDPE, HIPS and PVC are all equally important, low-density polythene dominates in flexibles with a European annual tonnage equal to all the others put together.

Plastics films may be made by casting or extruding sheet or by blowing a film, and the latter process is particularly important. The molten polymer is extruded through a ring-shaped die, and the tube so produced is blown into a tubular shape with air pressure. The bubble is flattened by pinch rolls and wound up ready to be made into bags, or side-slit to produce film. The width of the film and its thickness are controlled mainly by the dimensions of the die, the air pressure, and the extrusion and haul-off speeds.

3.4.6.4 Flexible laminates. A package for food has to satisfy several different criteria, and in many cases these cannot be met by a single flexible material. This has led to the development of a wide range of composite materials, made up of separate webs joined together to form laminates. The separate plies will contribute their own specific characteristics, such as stiffness and printability from paper, toughness from polyester, gas barrier from foil, or ease of heat-sealing from LDPE or EVA. The two main methods of forming laminates are by adhesion or co-extrusion. There are many variations and/or combinations of these two techniques, such as extrusion lamination, polythene coating of paper or board, metallisation, and so on.

Adhesive lamination. The individual plies are here joined with adhesives, which may be solvent- or water-based, or consist of molten waxes or polythene. Water-based adhesives like starch or casein need one permeable ply, preferably paper, to allow the escape of water vapour. Solvent-based adhesives of the polyurethane type are widely used, as they can join together virtually all the currently used materials.

One film is first coated with adhesive, and then passed through the drying oven to remove most of the solvent. The second film is then joined to the tacky surface of the ply emerging from the oven, by passage through combine rollers. So long as the process is controlled so as to give low solvent residuals, there should be no problems with food tainting.

Co-extrusion. In this process several different polymer streams are combined in a single die to form the laminate. Where the required polymers do not adhere easily to each other, thin adhesive polymer layers such as the acrylics are used as 'ties'. Five- or six-layer co-extrusions are commonplace, such as OPP/tie/EVOH/tie/LDPE, and a particularly important development is the range of pearlised films typified by Shorko P. The cavitated air-filled structure eliminates 'show-through' by foods such as chocolate, and also gives low densities (around 0.7 g/cm^3) with consequent higher yields. While co-extrusion avoids the high capital costs of adhesive laminating lines, its use is restricted to thermoplastic components.

Applications for laminates. While marketing and strength requirements will frequently rule out its use, nevertheless flexible packaging is the cheapest way to preserve most dried or liquid foods, including those that are aseptically packed. While on the borderline with respect to stiffness and three-dimensional aspects, the Tetra Pak is nevertheless a flexible package, since it incorporates the vertical form/fill/seal principle, whether for cleanfill or aseptic packaging applications. Table 3.5 lists some typical laminate structures used for various categories of food.

3.5 Heat-sealing

The thermoplasticity of plastics, mentioned previously, means that they are ideal as the heat-seal closing media for flexible laminate pouches, or for diaphragms on plastics, metal, glass or composite containers. The low-density polythenes and co-polymers are most widely used for this purpose, but the other polyolefins and coatings such as the acrylics, butyl methacrylate, or PVDC, are all used for specific heat-sealing applications. Cold-seal coatings, based on aqueous rubber latexes, are widely used for confectionery such as chocolate bars softened by heat. However,

Table 3.5 Flexible laminates for foods

Food	Properties required	Typical materials*
Frozen foods	Medium OTR/MVTR, flexibility	LDPE, PA/EVA, OPPc/p
Bacon, cured meats	Low OTR/MVTR, grease resistance	PVDC/OPP/LDPE, MXDT/LDPE
Cheese	Low OTR/MVTR, grease resistance	PA/tie/EVA, PVDC/OPP/LDPE
Meat and fish, fresh	Low OTR/MVTR, anti-mist	PA/ionomer, PET/PVDC/LDPE, EVA/EVOH/EVA
Milk, fresh	Medium MVTR, light barrier	LDPE/black LDPE, LDPE/board/ LDPE
UHT milk (aseptic)	Low OTR/MVTR, light barrier	LDPE/board/PE/Foil/PE/LDPE
Pasteurised foods	Low OTR/MVTR, heat resistance	PVDC/PET/MDPE, PP/foil/PP
Heat-sterilised foods	Low OTR/MVTR, high heat resistance	PET/PA/foil/EP
Wine, fruit juices	Low OTR/MVTR	MetPET/MetPET/EVA,LLDPE/tie/ EVOH/tie/LLDPE

* There are many equivalent specifications

they do not provide a seal of sufficiently high integrity for aseptic packaging purposes.

Heat-sealing is mainly carried out with electrically heated metal jaws, which are patterned to reduce the chance of poor surface contact. Heat may also be generated at the seal faces by the use of ultrasonically vibrating jaws, or by induction heating where there is aluminium foil in the laminated material. They may also be designed to 'squeegee' food contamination away from the seal area, via various designs of longitudinally interlocking jaws. When particulates are being filled the problem is more acute since some protein or vegetable tissues may be hard enough to remain trapped in the seal area. Hence every effort must be made to avoid seal contamination in the first place.

If two plastics faces are completely fused together, the seal will be completely impervious to all forms of contamination including microorganisms. There are many food packaging situations where an easy-open 'peel-seal' is required, and this may be obtained by modifying the seal conditions to prevent complete fusion, or by the use of dissimilar plastics for the two sealing faces. However, for aseptic packaging, where a fully fused seal is usually desirable, fusion will be achieved by control of the following sealing parameters.

1. Pressure. If this is inadequate the faces will not touch over the whole seal area. If it is too great the seal will be thinned and weakened.
2. Temperature. This must be high enough to melt the plastics faces as otherwise complete fusion cannot take place. However, if the temperature is too high the seal will again be thinned.
3. Time. When the temperature and pressure have been set correctly, sufficient time must be allowed for the heat to reach and melt the seal faces.

Maintenance of the established sealing parameters is particularly vital where form/fill/seal machinery running at high speeds is concerned, and so microprocessor control is essential. It is also pertinent to remember that a different supplier's material may not run without trouble unless new sealing parameters are established. This is not necessarily because the material is of poorer quality, but because minor variations in gauge or plastics melting points can affect seal performance, particularly with high-speed running.

3.5.1 Testing for leaking packs

The aim of total quality management, via procedures laid down in BS 5750 or similar standards, is that of '**zero defects**', so that faulty products will not reach the consumer. While the aim of 100% perfection leaving the factory gate may not be attainable, nevertheless very low failure rates are consistently achieved for aseptic lines that pack liquid products. A level of 1 in 10 000 maximum defectives or better is the normally accepted quality standard in the food industry.

While the highest standard of package integrity is achieved through control of the packaging process itself, this will always be supported by leak-testing procedures. These may be either on-line or carried out via batch sampling. Sensitive on-line tracer gas methods, such as the use of helium followed by mass spectroscopy detection, or sulphur hexafluoride and subsequent electron-capture detection, are feasible for empty containers. However, they cannot be employed for packs containing food, and so less-sensitive methods are used. Cans or metal-capped jars with internal vacua can be checked for vacuum levels by mechanical methods that measure lid concavity, or via electronic devices such as the Taptone, which relates to sound frequency. An on-line system has recently been developed for detecting leaks in flexible packs that have been gas-flushed with modified atmospheres such as CO_2/N_2. This works by measuring the resistance to an applied pressure, and is fully automatic at speeds up to 100 ppm.

Batch testing is always carried out in addition to on-line testing, and where aseptic packaging is concerned, this is likely to be the main safeguard against leakers. Package bursting, vacuum application to show leakage paths, dye wicking, challenge biotesting and other tests will enable production quality to be maintained provided that these procedures are coupled with close control over the in-plant packaging lines. The best way to prevent faulty products reaching the consumer is by 100% incubation of production, followed by inspection for leaking or blown packs. However, while this is essential for a new factory situation, or where a major change has been made, it should not be needed once a process has been fully established.

3.6 Shelf-life

The shelf-life of a packaged food is difficult to define with any great precision, because it is mainly a subjective judgement, dependent on the person tasting the food. Nevertheless the statement, 'shelf-life is that period before significant flavour change occurs', is an adequate description in general terms. It should be noted, however, that while the acceptability of a food depends mainly on the flavour, which is a combination of taste and smell, other factors also contribute to unacceptability. Solids may separate from liquid milks to give poor 'mouth feel', or colour changes may occur, such as browning of tomato ketchup.

At one time there was an accepted demarcation between fresh products such as milk, meat, fish, and so on, which were taken as short shelf-life foods, and heat-processed food in glass jars and cans, which could be kept indefinitely. However, modern marketing methods have encouraged the use of plastics packages, permeable to oxygen and moisture, which seldom provide indefinite shelf-lives. In addition, they have sought an extension of shelf-life for fresh fruits, fish, meats, dairy products, salads and so on, leading to the development of modified atmosphere packaging (MAP) techniques. Examples of modified atmospheres are the use of transparent laminates and carbon dioxide and nitrogen mixtures to extend the shelf-life of salads from 3–14 days, or cheese from 1–6 weeks. Also, a high-oxygen atmosphere, around 70%, plus carbon dioxide and nitrogen, is used to extend the shelf-life of red meats from 2–10 days.

3.7 Aseptic packs

The main cause of food deterioration and spoilage overall is microbiological, in other words by bacteria, moulds and yeasts. However, where aseptic packaging is concerned, the food processing has eliminated this factor and deterioration will be mainly oxidative, plus some inherent biochemical changes in the foods. In particular, UHT milks containing dairy lipids will develop stale and rancid flavours, and fruit juices will lose some of their aromatic flavour characteristic. Strongly flavoured soups and sauces are only affected by oxygen to a small degree, in spite of their animal and vegetable fat content.

The larger the amount of oxygen available to attack the product the greater will be the deterioration, and so packaging with the highest possible barrier to oxygen will be used for sensitive foods. However, it is equally important to ensure that the initial oxygen content of the packs is as low as possible, via minimum headspaces or nitrogen gas flushing. Tetra Pak and Combibloc both use foil laminates for long-life UHT milk to ensure a good oxygen barrier, and hence a shelf-life at ambient conditions of more than 6 months. Most foods are temperature-sensitive, and a shelf-life for UHT milks of 6 months in Northern European conditions would be 1–2 months in the tropics. Conversely, if the milks were held at chill conditions an extension of shelf-life would be obtained.

Aseptic dairy desserts or rice puddings in plastics pots do not have the protection of foil, although they will be made from high-barrier material and be closed with a foil diaphragm. The laminate material, typically PP/PVDC/PP, will also have internal dark pigmentation to reduce the catalytic effect of light on the oxidative processes. Sweetened desserts are not so sensitive to oxidation as whole milk, and so a shelf-life of more than 6 months is obtained even without the foil barrier.

3.8 The cost of packaging

The food industry is the largest industrial sector in Europe and in most countries around the world whether fully developed or not. In the developed countries, however, packaging has also grown with the food industry to become very big business, measured either in tonnes of materials or value in monetary terms. In the UK for example, the 1989 Consumer Expenditure Survey shows expenditure on food and drink to be £59 billion, which averages 18% of disposable income after tax. The cost of food and drink packaging is not available in these statistics, but it is estimated to be around £3.8 billion, in other words, 6.5% of the retail price of the food and drink.

While the average cost of food and drink packaging, including outer packaging, is given as 6.5%, there will naturally be large variations in this figure depending on the foods being packed. A Kraft paper or polythene bag for fresh fruit will be at the bottom of the cost scale, with high-barrier processable plastics containers or drawn aluminium cans coming near the top. These latter containers could well be at a level of 15–20% of the retail value of the food product. However, it should be noted that the cost of the packaging operation itself, where there is a food heat-preservation step combined with it, such as heat preservation or aseptic filling, could bring the total packaging costs up to around 25%.

3.8.1 Cost-effective packaging

The dictionary definition of cost-effective is 'giving adequate return for outlay', and while this is easy to state it is far more difficult to achieve in practice. Such a pack will not necessarily be either the cheapest or the dearest, but it will be the least expensive one that can perform adequately all the functions required of it.

The primary task of the packaging must be to preserve the food, under all the conditions it will encounter for the shelf-life envisaged. However, the package must also run smoothly on the packaging lines with minimum down time, be handled and opened easily, and not present disposal problems at the end of its life. Apart from these technical requirements, it also has a most important role in the marketing of the food products.

With regard to the materials competing for the food-packaging business, comparative usage on a monetary basis is likely to remain relatively steady during the 1990s. In the UK, paper and board is the largest sector at 47%, metal and plastics are 22% each, with glass at 6%, and miscellaneous materials at 3%. While composite figures for Europe are not available, it would appear that they are somewhat similar to UK percentages but with a greater expenditure on plastics packaging. Taken over the long term, price trends are much the same for all the materials, around the level of inflation, and there is no immediate prospect of one material becoming suddenly more cost-effective than another.

3.9 Food-packaging legislation

The increase in the production of packaged food has led to the worldwide enactment by governments of legislation to control the wholesomeness of food packaging. National organisations, such as the USA Food and Drugs Administration, the BGA German Federal Health Office, and the UK's Ministry of Agriculture Fisheries and Food, all issue regulations. In addition Europe also has supranational regulations, and there are also many codes of practice that are not legally binding. The regulations are broadly in two areas, trading standards and materials safety. Trading standards will encompass aspects such as sizes of pack and labelling requirements, for example, net weight, use-by dates, and so on. Materials safety seeks to ensure that the health of consumers is not put at risk by the packaging.

However, following concern over the use of natural resources and the disposal of waste packaging, much of the new legislation worldwide concerns environmental rather than food-safety issues. The UK's Environment Protection Act 1990 already lays down strict guidelines for waste disposal and litter prevention by local councils, and the EC's Directive on Packaging and Packaging Waste has far-reaching implications for European industry.

3.9.1 EC legislation

Since 1985, the European Commission has taken the lead in promoting food-packaging legislation in Europe as a whole, and this policy has been greatly helped by the adoption of the Single European Act, which allows agreement by majority. However, it is unlikely that full harmonisation of EC legislation will be complete by the end of 1992. One of the reasons for this is that, whereas some member states such as France operate Statute (Roman) Law, the UK has a Common Law system, and Europe as a whole is a mixture of the two. While Common Law gives general guidance and works to precedents, Statute Law lays down specific requirements in detail.

3.9.1.1 Current EC legislation. The legislation currently arising from the European Community (EC) may be grouped as follows:

1. Framework Directive 89/109: General rules for all materials and articles intended to come into contact with food;
2. Directive 80/590: Symbol for use on food packaging materials;
3. Directive 83/229: Composition of regenerated cellulose films (RCFs);
4. Directive 86/338: DEG/MEG limits in RCFs (lower limits proposed in new draft Directive);

5. Directive 84/500: Ceramics—migration limits for lead and cadmium;
6. Directive 90/128: Plastics—outline of overall migration-testing methods and limits, and a positive list of monomers;
7. Directive 78/142: Vinyl chloride monomer limits in packaging and food;
8. Directive 80/766: Method of analysis of VCM in materials;
9. Directive 81/432: Method of analysis of VCM in food;
10. Directive 82/711: Migration-testing procedures—basic rules, i.e. definition of plastics, simulants, times, temperatures; and
11. Directive 85/572: Migration-testing simulants—classification of foods and reduction factors.

In the field of plastics, work is well advanced (1991) on positive lists of additives and manufacturing residues such as processing aids. Other packaging materials have Directives in preparation, for example glass, metals and alloys, paper and board, wood, rubbers, waxes, and so on.

3.10 Packaging and the environment

For the foreseeable future, it is likely that the question of the environment will be the most pressing problem facing the packaging industry worldwide. Part of the difficulty arises from the fact that the public's assessment of packaging is that, at worst, it is expensive and not needed and, at best, a necessary evil. In addition, there are problems with litter, which give a high profile to an undesirable aspect of packaging. Both these facts mean that politicians at national or local levels, encouraged by media stories, tend to introduce legislation that is not particularly helpful; for example, the beverage container ban in Denmark and USA State deposit laws, which all shift the problem elsewhere. For a long time the packaging industry worldwide hindered the development of rational solutions to its problems by pretending they did not exist, and over-reacting to criticism. However, a new spirit of cooperation between government and industry now augers well for the future.

The four main environmental questions that require to be addressed are: (i) litter; (ii) waste disposal; (iii) natural resources; and (iv) recycling. All four are interlinked to some extent; however, it should be noted that, separately or collectively, these issues are very difficult and complex. Certainly no complete solutions will ever be found unless government, industry and consumers cooperate to ensure that codes of practice and legislation work efficiently. In brief, the current situation is as follows.

3.10.1 Litter

The prevalence of litter in public places is primarily a social problem, which biodegradable packaging certainly could not solve. In countries where on-the-spot fines are unacceptable there must be a concerted campaign of public education. Studies have shown that the proportion of 'litter louts' in a population is usually quite small, since most people will not scatter rubbish if there is none already on the ground. Hence a comparatively small effort by local authority employees to keep areas clean can go a long way towards solving the problem. It goes without saying that there must also be sufficient, conveniently placed, waste receptacles for the public to use.

3.10.2 Disposal of waste

After rubbish has been collected, the two most widely used ways of disposing of it are by landfill and incineration.

3.10.2.1 Landfill. This is always the most popular method wherever there are suitable sites near enough to the waste collection centres to make transport costs economic. Apart from aesthetic drawbacks, the practice of tipping is quite logical, provided the sites are properly compacted and no toxic chemicals are illegally introduced. As a whole, about 75% of the EC's domestic waste is disposed of by landfill, with the highest proportion in the UK at 90%, followed by Germany 75%, France 70%, and Denmark 40%. Owing to an acute shortage of suitable terrain, Japan uses landfill for less than 10% of its waste.

The concern often expressed that plastics destabilise tips has recently been shown by Professor Rathje in the USA to be unfounded. His investigations indicated that the percentage of plastics by weight was only about 12%, rather than the usually quoted 30%. At this level, plastics compacted very well and the tips were completely stable. Another advantage was that explosive gases such as methane were not trapped by plastics containers.

3.10.2.2 Incineration. This technique is largely made feasible by plastics packaging, because polypropylene and the polythenes have a calorific value 50% higher than coal. If this were not so, fuel would need to be added to the waste to enable it to burn. While incinerating plants are expensive, some of the cost can be recouped by generating electricity for heating and lighting.

It has been said that the hydrochloric acid (HCl) evolved when PVC packaging is incinerated is a major contribution to 'acid rain', but in Europe at least, 75–80% of the HCl in acid rain comes from power stations. Again, the possible formation of harmful dioxins through the incineration of chlorine-containing plastics has been used to condemn incineration techniques, although the high temperatures at which modern incinerators operate virtually precludes dioxin formation.

3.10.3 Use of natural resources

While all the world's natural resources and non-renewable fossil fuels in particular should be used wisely, packaging materials are not generally excessive users of these valuable assets. Paper for books, magazines, stationery and packaging comes mainly from softwoods, which are being replaced as fast as they are used. The deposits of sand, soda ash, limestone, and other chemicals from which glass is made are inexhaustible, as are the iron and aluminium deposits needed to make metal cans.

Plastics are often quoted as users of dwindling oil reserves, but they only consume around 5% of the world's oil with about half this plastics packaging. Also, as previously mentioned, some of the energy used to make plastics is recovered via incineration techniques.

3.10.3.1 Energy conservation. Within the packaging industry itself there is often discussion as to which packaging material, of glass, plastics, steel, paper or aluminium, is the most energy efficient. Since the bulk of energy usage is for heating purposes this is not a very important environmental issue. However, comparison of beverages containers on a volume basis shows an aluminium can to use more energy than a steel can, than a one-way glass bottle. Returnable glass bottles are even more energy efficient, provided that they are returned!

3.10.4 Recycling

There is currently a strong consensus of opinion that the packaging industry can best help to protect the world's resources by recycling as much as possible. Not only are raw materials saved but energy too, since the primary energy used to produce them has already been discounted.

The EC Commission's recent Directive on *Packaging and Packaging Waste* proposes to tackle the problem as a whole, rather than concentrating on specific areas such as beverage containers or PVC. The legislation will cover waste from industrial processing, transit packaging, agriculture, shops, and so on, as well as household rubbish. There are proposals for reducing the amount of packaging used, preventing refillable products moving to one-way packaging, banning unnecessary secondary packaging, and for the imposition of deposit schemes and taxes. However, the central theme of the proposed laws is that the amount of packaging recycled must be increased to a level of 60% within 5 years of the Directive becoming law.

3.10.4.1 Current recycling. Table 3.6 gives the estimated 1991 figures for recycled packaging materials in Europe. These figures would be much higher if all recycling was economically feasible; however, money will have to be found when the EC Directive becomes law.

Table 3.6 European recycling levels in 1991

Paper and board	55%
Glass	35%
Metal	20%
Plastics	15%

It should be noted that around half the metal figure and three-quarters of the plastics is made up of re-use of factory skeletal waste. Hence there is no need for this to be cleaned of stale food residuals and other contaminants, which so often makes recycling uneconomic. The recycling of paper and board and glass is very well established because there are sound financial reasons for doing it. In addition, all the necessary technology is available, even for processing the polycoated board foil laminates used for aseptically packed milk and other beverages.

3.10.4.2 Metal recycling. Tinplate container recycling has long benefited from the availability of magnetic separation from household waste. Now a recently developed rapid detinning process allows high-quality steel to be made from non-incinerated waste, and also produces tin ingots. Europe's first purpose-built plant for sorting aluminium and steel cans for recycling was opened in Swansea in 1989, and more are planned. While only 15% of beverage cans are recycled in Europe, the USA figure is now more than 60%.

3.10.4.3 Plastics recycling. Apart from in-plant situations, the recycling rates for plastics are low because of the particular problems they pose. There are so many different types that sorting is a major problem, even though the marking of containers to identify the material is gaining ground. Additionally, they all absorb aromas and vapours to some extent which makes cleaning difficult. Finally, in most cases, their low cost and weight tends to make collection economically unattractive. The success of PET bottle recycling schemes has stemmed largely from the high initial cost of the material, and from the fact that a non-food textile market was available for the recycled polymer. The USA PET recycling level is now over 25%.

Although recycling is an important technique, it should not be looked at in isolation as a universal answer to the environmental problems of packaging. The proposed EC *Waste* Directive sensibly requires packaging to be assessed for environmental impact over its whole life cycle, in other words by a 'cradle-to-grave' analysis from raw materials to waste disposal. Hence for example, if the energy and industrial resources needed to recycle say a particular plastics pack were excessive, then incineration to produce energy could be a better option. In addition the EC's 'eco-labelling' system, contained in the Directive, will only approve those packs that satisfy such an analysis. This logical approach to environmental impact ensures that the most suitable method of waste disposal will be chosen, and hence that no aseptic or other food pack need be banned solely because it did not fit into a particular recycling scheme.

4 Ohmic heating

P. J. SKUDDER

4.1 Introduction

4.1.1 Consumer products

The past decade has seen a tremendous growth in the amount and variety of ready prepared food products available to the consumer. At present, the demand is satisfied, to a large extent, by cook-chill processing where improved technology has resulted in a wider range of products and increased quality, compared with traditional frozen and canned food.

However, cook-chill food has only a very limited shelf-life due to the presence of microorganisms and must be stored under strictly controlled refrigerated conditions. In-container sterilisation enables food to be stored at ambient temperatures for long periods, but the quality may be poor because of the need for extensive peripheral heating to ensure sterilisation of the food at the centre of the container. This is especially so with large cans used for catering purposes.

4.1.2 Aseptic processing

The alternative is aseptic processing of the food using plate, tubular or scraped surface heat exchangers followed by filling into paperboard or plastic containers under aseptic conditions. Aseptic processing has been used commercially for over 30 years for milk and fruit juices and has been widely adopted by the food industry.

However, for products containing large particulates, up to 25 mm cubes, the use of conventional heat transfer techniques necessitates over-processing of the liquid phase to ensure that the centre of the individual particles is sterilised. This can result in destruction of flavours and nutrients, and particle damage.

4.1.3 Ohmic heating

These heat transfer problems have now been overcome with the development of ohmic heating, which operates by the direct passage of electrical current through the continuous flow of food product. Heat penetration throughout the product is far more rapid and even, resulting in high levels of flavour retention and particulate integrity, previously unrealised.

When combined with an aseptic container or bag-in-box filling system, this process is capable of producing high added-value ready prepared meal products with a quality equal to cook-chill but with the benefit of safe, long-term storage under ambient conditions and without the need for costly refrigeration. The system can also be applied to fruit products, such as whole strawberries and diced fruits.

Ohmic heating systems have now been in commercial operation in the UK for over 2 years. A plant has recently been installed in Japan and further plants will shortly be installed in France, Japan and the USA processing a variety of low- and high-acid products.

4.2 Principles of aseptic processing

In order to appreciate the benefits of ohmic heating, it is necessary to understand the basic principles of thermal sterilisation technology.

4.2.1 Thermal sterilisation—the options

Ambient stable products generally depend on thermal processing to inactivate microorganisms and render the product commercially sterile. The definition of commercial sterility is a product in which microorganisms may be present but conditions are such that they are not allowed to grow. Any microorganisms present must not cause spoilage of the food during the stated shelf-life nor must they be of concern to the health of a consumer.

This can be carried out in two ways:

1. In-container sterilisation, where the product is sterilised inside sealed containers, such as cans, glass jars and plastic pots or trays.
2. Aseptic sterilisation, where the product and package are sterilised separately and brought together for filling and sealing in a sterile environment.

4.2.2 Principles of thermal sterilisation

There are several basic principles which apply to thermal sterilisation of food products. Food products are classified into two broad classifications: low-acid and high-acid products. The extent of the heat treatment given to a low-acid product where the pH is greater than 4.5 is given by the F_0 value. The F_0 value is defined as the total integrated lethal effect expressed in minutes at a temperature of 121.1°C for microorganisms that have a Z value of 10°C. The Z value is the temperature change that gives a 10-fold change in the decimal reduction time. The calculation of an F_0 value is given as:

$$F_0 = \int L \, dt \tag{4.1}$$

where L is the lethality (i.e. the number of minutes at the reference temperature equivalent to 1 min at the actual temperature applied for sterilisation). The lethality is calculated from:

$$L = 10^{[(T - 121.1)/10]} \tag{4.2}$$

Low-acid food products are capable of sustaining the growth of *Clostridium botulinum* and must be heat-treated to the equivalent of at least 121.1°C for 3 min (an F_0 value of 3) in order to achieve a 12 decimal reduction of the microorganism. Depending on the actual pH and nature of the food, it is commonly necessary to apply even higher F_0 values (e.g. F_0 8 to 20), as some food spoilage organisms are considerably more heat-resistant than *Clostridium botulinum.*

High-acid products are those with a pH of less than 4.0, typically fruit products, and will not normally allow the growth of food-poisoning microorganisms. These are typically heat-treated at 90–95°C for a period of 30–90 s, to inactivate yeasts, moulds and *Lactobacillus* organisms.

The above process conditions must, however, be applied to the slowest heating point in the product, which is usually in the centre of the largest particle situated at the centre of a can. It is for this reason that traditional in-container sterilisation may produce poor-quality products, because of the need for extensive heating to ensure sufficient heat penetration to the centre of the container. Such a heat treatment necessarily results in substantial over-heating of the liquid phase, with consequent over-heating of the external surfaces of the particulates.

4.3 Processing options

Continuous aseptic processing implies continuous flow of the food product. The choice of equipment to be used is initially determined by the physical characteristics of the product under flow conditions. Several systems are commonly used commercially.

4.3.1 Indirect plate heat exchangers

These are generally used for homogeneous, low-viscosity fluids containing no suspended material, for example milk, juice and thin sauces.

4.3.2 Indirect tubular heat exchangers

These are generally used for homogeneous higher-viscosity liquids and those containing particles up to approximately 10 mm; for example soups, sauces and fruit purées.

4.3.3 Direct steam injection and steam infusion

This is used for higher-viscosity, homogeneous liquids and is particularly suited for shear-sensitive products and those that may be susceptible to fouling, for example creams, desserts and viscous sauces.

4.3.4 Scraped surface heat exchangers

These are used for heavy viscous liquids or those containing particles up to approximately 15 mm in diameter. Solids loadings of up to approximately 40% w/w can be processed, for example diced fruit preserves, purées and soups.

4.4 Ohmic heating

4.4.1 Principle of ohmic heating

The ohmic heating effect occurs when an electric current is passed through an electrically conducting product (Figure 4.1). In practice, low-frequency alternating current (50 or 60 Hz) from the public mains supply is used.

In common with microwave heating, electrical energy is transformed into thermal energy. However, unlike microwave heating, the depth of penetration is virtually unlimited and the extent of heating is governed by the spatial uniformity of electrical conductivity throughout the product and its residence time in the heater.

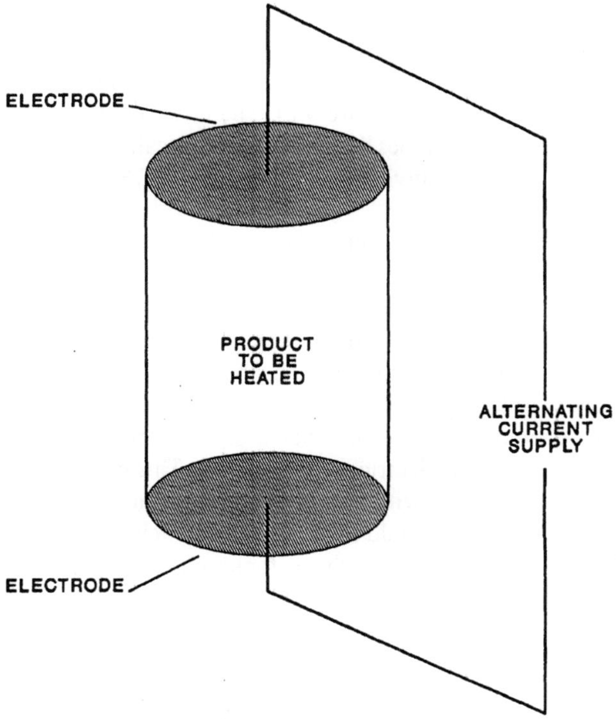

Figure 4.1 Principle of ohmic heating. An electric current is passed through an electrically conducting product.

4.4.2 Advantages of ohmic heating

For most practical purposes, the product does not experience a large temperature gradient within itself as it heats, and liquid and particulate are heated virtually simultaneously. The requirement to over-process the liquid to ensure sterility at the centre of a large particulate (as with scraped surface and tubular heat exchangers) is therefore reduced. This results in less heat damage to the liquid phase and prevents over-cooking of the outside of the particulates.

Another major advantage is that there are no hot heat transfer surfaces, which reduces the possibility of deposit formation and the transfer of burnt particles to the finished product. Nor is there a need for mechanical agitation of the product, which can cause loss of true particulate identity. The applicability of ohmic heating is dependent on product electrical conductivity. Most food preparations contain a moderate percentage of free water with dissolved ionic salts and hence conduct sufficiently well for the ohmic effect to be applied.

4.5 Design of the ohmic heater

4.5.1 Mechanical design

The ohmic heater column typically consists of four or more electrode housings machined from a solid block of PTFE and encased in stainless steel, each containing a single cantilever electrode (Figure 4.2). The electrode housings are connected using stainless steel spacer tubes lined with an electrically insulating liner. Suitable lining materials include polyvinylidene fluoride (PVDF), polyether ether ketone (PEEK) or glass.

The column is mounted in a vertical or near-vertical position with the flow of product in an upwards direction. A vent valve positioned at the top of the heater ensures the column is always full. The column is configured such that each heating section has the same electrical impedance and hence the interconnecting tubes generally increase in length towards the outlet. This is because the electrical conductivity of food products usually increases with increase in temperature; indeed, for aqueous solutions of ionised salts, there is a linear relationship between temperature and electrical conductivity. This phenomenon is attributed to increased ionic mobility with increase in temperature and also applies to most food products. Exceptions could be products in which viscosity increases markedly at higher temperatures, such as those containing previously ungelled starches.

4.5.2 Temperature control

Commercial ohmic heating plants are available with a fully automatic temperature control system. Inlet changes that will affect the final product outlet temperature

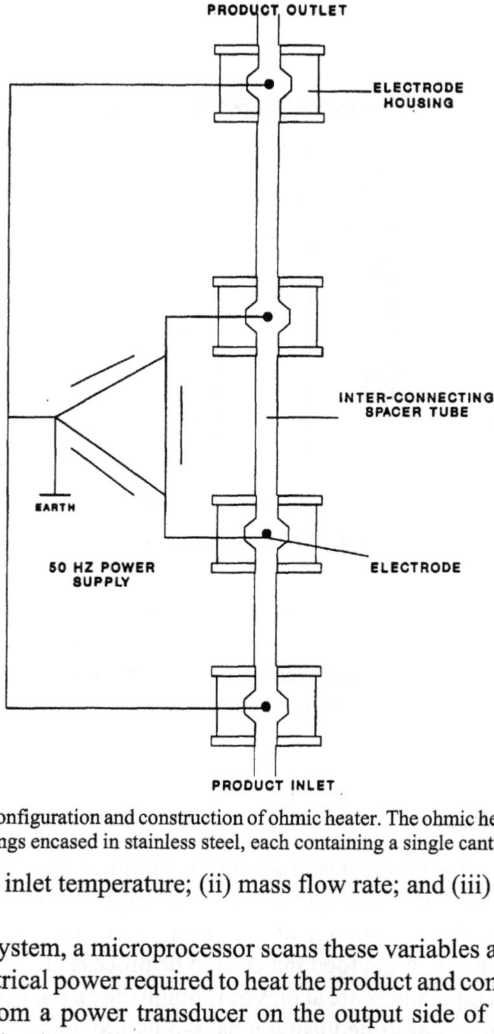

Figure 4.2 Electrical configuration and construction of ohmic heater. The ohmic heater column consists of electrode housings encased in stainless steel, each containing a single cantilever electrode.

are: (i) changes in inlet temperature; (ii) mass flow rate; and (iii) product specific heat capacity.

In the control system, a microprocessor scans these variables and continuously computes the electrical power required to heat the product and compares this value with the signal from a power transducer on the output side of the transformer. Feedback monitoring is used to prevent any long-term drift in outlet temperature.

4.6 Aseptic processing using the ohmic heater

4.6.1 Processing system

The ohmic heater assembly can be incorporated into a complete product sterilisation or cooking process. A typical line diagram of such a plant is shown in Figure 4.3. In common with other types of continuous system, careful design of the ancillary process equipment is necessary and, once heated, the product must be cooled by more conventional means using scraped surface or tubular heat exchang-

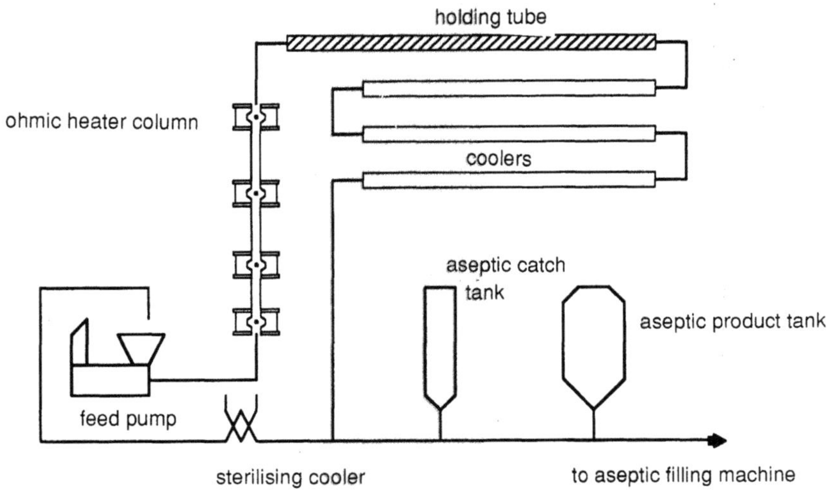

Figure 4.3 Line diagram for aseptic processing using the ohmic heater.

ers. The latter are generally preferred for particulate processing in order to maximise the greatest advantage of ohmic heating, that of a minimal structural damage to the particulates.

4.6.2 Plant sterilisation

Pre-sterilisation of the ohmic heater assembly, holding tube and coolers is carried out by recirculation of a solution of sodium sulphate or an alternative food-compatible mineral salt solution at a concentration that approximates to the electrical conductivity of the food material to be processed.

Sterilisation temperatures are achieved through the passage of electrical current and back pressure is controlled using a back-pressure valve. The aseptic storage reservoir, interface catch tank and connecting pipework to the filler are traditionally sterilised using pressurised steam.

The use of a sterilising solution of similar product electrical conductivity minimises adjustment of electrical power during subsequent change to product, thus ensuring a smooth and efficient changeover period with little temperature fluctuation.

4.6.3 Product processing

Once the plant has been sterilised, the recycled solution is cooled using a plate heat exchanger in the recirculation line. When steady-state conditions are reached, sterilising solution is collected or run to drain, and product is introduced into the hopper of a positive displacement feed pump. Typically, this would be a Marlen twin-piston metering pump.

Back pressure during the changeover period is controlled by regulation of top pressure in a interface catch tank using sterile compressed air or nitrogen. This tank serves to collect the sodium sulphate/product interface. Once the interface has been collected, product is diverted to the main aseptic storage vessel where top pressure is similarly used to control back pressure in the system. Back pressure is maintained at a nominal 2 bar when heating high-acid food products at temperatures of 90–95°C. A 4 bar back pressure is used for low-acid food products where sterilisation temperatures of 120°–140°C are applied. This higher back pressure is necessary to prevent boiling of the product in the ohmic heater system.

Safety features are incorporated to ensure power is automatically switched off should there be any loss of pressure.

The use of pressurised storage vessels has proved to be an extremely effective method of controlling back pressure in ohmic heaters where throughputs are less than 750 kg/h. In larger systems operating in excess of 2 t/h, an aseptic positive-displacement pump downstream of the cooler can be used as an alternative, depending on the composition of the food product. This alternative overcomes the requirement for two aseptic storage vessels in order for the system to continuously feed product to the filling machine.

Product is progressively heated to the required sterilisation temperature as it rises through the ohmic heater assembly. It then enters an insulated holding tube before being cooled in a series of tubular heat exchangers. After cooling, product enters the main storage reservoir prior to aseptic filling.

4.6.4 Alternative cooling

More rapid cooling can be achieved with some products in which the final particulate level will be less than 40%. This system involves a combination of ohmic and traditional heat treatment and takes advantage of the capability of the ohmic heater in being able to process products containing up to 80% particulates.

During preparation, the product is formulated into a high concentration particulate stream and a separate liquid stream. The liquid stream is conventionally sterilised and cooled in a plate or tubular heat exchanger system before being injected into the particulate stream leaving the holding tube of the ohmic heater. Such an operation has the advantage that it reduces the capital and operating cost for a given throughput.

4.6.5 Cleaning

When food processing is complete, power is turned off and the plant rinsed with water. This is followed by cleaning using a 2% (w/v) solution of caustic soda, which is recirculated at 80°C for 30 min.

Owing to the high electrical conductivity of the cleaning solution, it is heated using a plate heat exchanger in the recirculation line, and not using the ohmic heating principle.

4.6.6 Throughputs

A range of ohmic heaters incorporating different sizes of electrode housings and interconnecting spacer tubes is commercially available to process food products at throughputs of up to 3 t/h, depending on the required temperature rise. A 5 kW laboratory system that is capable of a throughput of 50 kg/h has also been designed for research and development purposes.

4.7 Product quality

Included in the list of perceived advantages of continuous aseptic processing (Table 4.1) is that of improved product quality, with the basic standard being set by products sterilised by traditional canning methods. These quality standards include microbiological safety (i.e. process lethality), cooking effects, and nutrient/vitamin retention.

Unlike other processes requiring heat penetration, either from the outside to the centre of a can (in-can sterilisation) or from the outside of the particulate to the centre (tubular or scraped surface heat exchanger), ohmic heating has the ability to provide exceptionally rapid uniform heating of the product virtually simultaneously. This provides a high degree of microbiological security, as there is a very narrow range of lethality from the surface to the centre of the particulates.

4.7.1 Microbiological process validation

The ohmic heating process has been validated in a practical operating situation using a technique developed by the Campden Food and Drink Research Association. In this technique, highly heat-resistant spores of a test organism are inoculated into alginate analogues of the food particulates, as described by Brown *et al.* (1984).

Table 4.1 Advantages of ohmic heating

Fresher tasting, more nutritional products can be produced containing large particulates
Ability to produce high added-value recipe dishes
Ability to heat food products in continuous flow without the need for any hot heat transfer surfaces
Process is ideal for shear-sensitive products
Heat can be generated in the product solids without reliance on thermal conductivity through liquid
System is quiet in operation
Low maintenance costs
Process is easy to control and can be started or shut down instantly
Potential savings in processing and packaging costs
Wide range of packaging options

The inoculated alginate particulates are added to the food product, which is then processed in the ohmic heater in the normal way. After processing, the particulates are recovered and the number of surviving spores is determined. This information, together with a knowledge of the heat resistance of the inoculated spores, enables the heat treatment given to the particulates to be described as equivalent to a certain number of minutes at 121.1°C (F_0 value).

The technique has been successfully applied to a wide variety of particulate food products, clearly demonstrating the beneficial volumetric heating effect that occurs during ohmic heating.

4.7.1.1 Overall F_0 *value for beef and carrot particles.* For example, with a product consisting of beef and carrot particulates in gravy, alginate analogues were prepared of both types of particulate by maceration and addition of sodium alginate. A known quantity of *Bacillus stearothermophilus* Th24 spores were added to the solution, which was formed into cubes by immersion in calcium chloride. The cubes were added to the feedstock and processed using a 45 kW ohmic heating plant under conditions calculated to give a theoretical F_0 of 32 min in the liquid phase. After processing, the inoculated cubes were recovered and the number of surviving spores determined (Table 4.2). The values received by the carrot particulates ranged from 28.1–38.5 while the meat particles received an F_0 of 23.5–30.5.

This experiment clearly illustrated that the particulates were heated directly by electrical resistance heating to a similar level as that received by the liquid phase. If the particulates had been heated by conventional heat transfer from the liquid, as in a tubular or scraped surface heat exchanger, then the theoretical F_0 value would have been only 0.2 at the centre.

From the narrow range of F_0 values received by the inoculated particulates, it is also evident that there was not a large difference in the residence time distribution of the particulates throughout the ohmic system. This is yet another advantage of being able to directly heat particulates rather than relying on heat transfer from the liquid phase.

This technique also enables very high concentrations of particulates to be processed, which greatly minimises the possibility of particulates either settling out or passing through the system at a faster rate than others.

Table 4.2 Results of lethality challenge tests during ohmic heating using spores of *Bacillus stearothermophilus* embedded in alginate particles

Product	Alginate particle type	F_0 range	Mean F_0 value
Beef and carrots in gravy	Beef (19 mm cube)	23.5–30.5	27.0
	Carrot (19 mm cube)	28.1–38.5	33.7
Beef and carrots in gravy	Beef (19 mm cube)	28.0–38.5	32.5
	Beef (3 mm spore bead inside 19 mm cube)	34.0–37.5	37.0
	Carrot (19 mm cube)	23.1–44.0	35.6
	Carrot (3 mm spore bead inside 19 mm cube)	30.8–40.2	37.1

4.7.1.2 F_0 *distribution for beef and carrot particles.* In a further experiment, the lethality received throughout the particulates was compared with that received at the centre. This comparison was made by preparing some alginate analogues as described above and others by setting half of the cube before adding a known amount of spores contained in a small alginate bead at the centre and forming the remainder of the cube.

With carrots, the F_0 received throughout the particulates ranged from 23.1–44.0 compared with 30.8–40.2 at the centre. The beef particulates received an F_0 of 28.0–38.5 throughout with an F_0 of 34.0–37.5 at the centre. This experiment indicates that the heat treatment given to the centre of the particulates is similar to that received throughout, with no significant over or under heat treatment of either the centre or outside edges.

4.7.1.3 Validation of commercial systems. It is recommended that this type of validation experiment is conducted with each type of food product to be commercially processed in order to establish the correct time/temperature conditions to ensure a safe, stable product is produced.

4.7.2 Cooking value

The cook value (C_0) of a process is a measure of the extent of cooking and vitamin destruction that occurs during heating. It is calculated in a manner similar to that

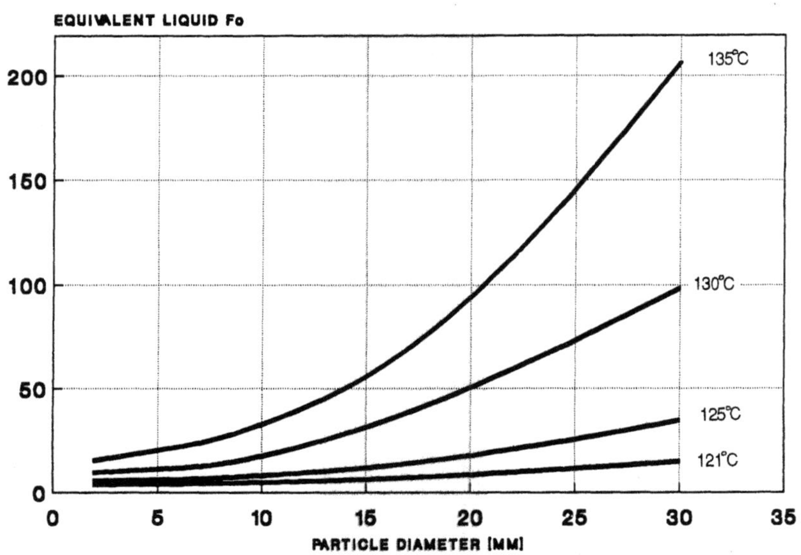

Figure 4.4 Effect of particle size on the F_0 value received by the liquid phase of a particulate-containing food in order to attain an F_0 of 5 at the centre of a particulate for sterilisation temperatures of 120–135°C, by conventional heat transfer.

for an F_0 value, except that the reference temperature is 100°C instead of 121.1°C, and the Z value is higher. This is because the rate of increase in heat inactivation of microorganisms with increase in temperature is greater than the rate of increase of cooking reactions and vitamin destruction. For example, the Z value for the heat denaturation of the vitamin thiamine is 33°C instead of the 10°C value for microorganisms.

In order to minimise cooking reactions it is, therefore, beneficial to heat the food at the highest possible sterilisation temperature at the highest possible rate.

However, with conventional heating it is just not possible to sterilise large particulate-containing foods at temperatures much above 130°C without serious over-heat treatment of the liquid phase. Figure 4.4 indicates the F_0 value received by the liquid phase in order that the centre of the particulates receives a F_0 value of only 5, as a function of the sterilisation temperature for different sizes of particulate. If sterilisation of a 25 mm particulate is attempted at 135°C then the liquid phase would receive an F_0 of nearly 150, which is a massive over-heat treatment. However, by restricting the particle size to 15 mm with a sterilisation temperature of 130°C, the liquid phase will receive an acceptable F_0 value of just over 25. These are the process conditions commonly employed for scraped-surface heat exchangers when processing low-acid food products.

Figure 4.4 also indicates why temperatures of 125°C are usually not exceeded with in-can sterilisation, especially when large particulates are present.

The additional constraint with conventional heat treatment is the time taken for the heat to penetrate to the centre of the particulate. The relationship between time taken for the centre of a particulate to reach an F_0 of 5 as a function of carrier liquid temperature and particle size is given in Figure 4.5. Any attempt to use conventional

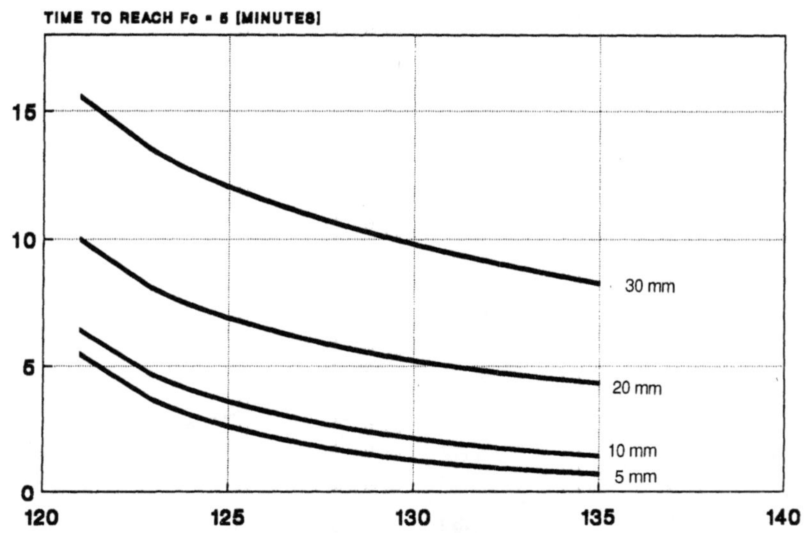

Figure 4.5 The effect of particle size on the time taken to attain $F_0 = 5$ at the centre of a particulate, by conventional heat transfer.

continuous heat treatment with particle sizes above 15 mm would necessitate the use of a holding tube to hold product for greater than 5 min. Such a tube would be excessively long.

Another consideration with conventional continuous heat treatment is that the concentration of particulates must generally be limited to about 30–40% in order that there is sufficient hot liquid available to heat the particulates. In addition, the viscosity of the carrier phase must also be limited because as the viscosity increases then so does the resistance to heat transfer from liquid to particulate (the film-temperature coefficient). Such considerations do not assist in maintaining particulates in suspension throughout the heating process, with the result that there can be large differences in the residence-time distributions between liquid and particulates.

With ohmic heating, however, the particulates are heated directly and sterilisation temperatures can be increased to 140°C (the operating temperature limit of the plastics components in the ohmic heater) without any such over-heat treatment of the liquid phase. More importantly, the ohmic heater is able to process much higher particulate loadings suspended in highly viscous carrier liquids, which greatly assists in minimising differences in residence time distribution.

The ohmic heater can, therefore, be used to sterilise large particulate food products at higher sterilisation temperatures while minimising the level of cooking and destruction of vitamins. The relationship between F_0 and C_0 value (based on the destruction of thiamine) as a function of sterilisation temperature is given in Figure 4.6. It can be seen that at 140°C an F_0 value of 24 results in a C_0 value of 4, while at 130°C an F_0 value of only 8 gives double this cook value.

In conclusion, the ohmic heater enables higher sterilisation values to be given

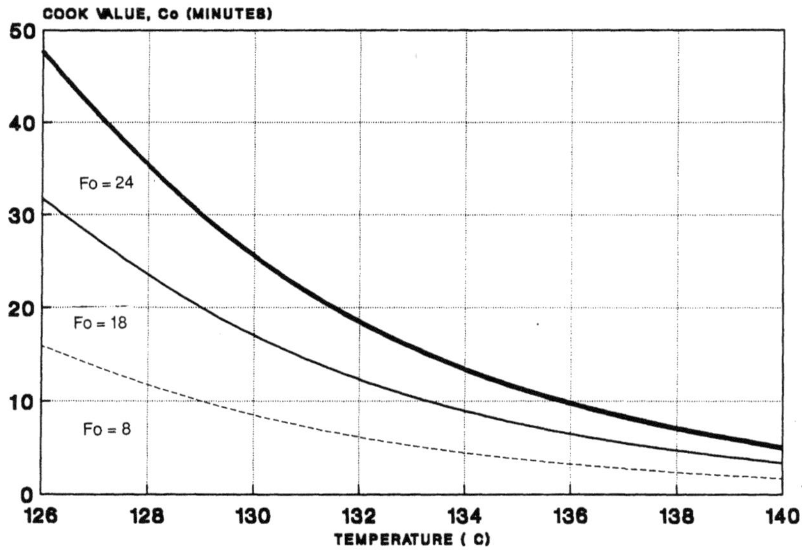

Figure 4.6 The effect of process temperature on the cook value (C_0 received by a particulate for F_0 values of 8, 16 and 24 min (based on Z value of 33°C).

Table 4.3 High-quality added-value products

High-acid products	
Vegetable	Ratatouille
	Pasta sauce with vegetables
	Vegetables provencale
Fruits	Fruit compote
	Strawberries
	Apple sauce
	Sliced kiwi fruit
Low-acid products	
Pasta	Tortellini in tomato sauce
	Cappaletti in basil sauce
	Tagliatelle à la crème
Meat	Beef bourguignone
	Beijing lamb
	Beef and vegetable stew
	Lamb wala gosht
Vegetable	Vegetable curry
	Minestrone soup concentrate

at lower cook values to large particulate food products with minimal differences in residence-time distribution. These are important safety benefits that cannot be ignored in selection of a process to produce safe high-quality ambient-stable food products containing large particulates.

4.8 Products

Typical exampes of such high-quality added-value products are shown in Table 4.3.

4.9 Commercial installations

Commercial ohmic heating installations are shown in Table 4.4.

4.9.1 Research systems

4.9.1.1 Continuous flow systems. These are 5 kW systems which have been designed to enable the food processor to carry out trials for optimisation of recipes and process conditions in a research and development environment, before production on a commercial system. They are listed in Table 4.5.

4.9.1.2 Batch systems. These are also 5 kW systems but enable the heating of a nominal batch of 1 kg product in a small ohmic heater cell. They are primarily designed for the measurement of the individual heating rates of particulates by inserting thermocouples into various components of the food product. Two such systems are currently under development for customers in Japan.

Table 4.4 Commercial installations

Plant	Customer	Location	Products	Filling	Throughput
300 kW Aseptic system	Confidential	Confidential	Low-acid particulate		Capable of processing up to approximately 3 t/h.
75 kW Aseptic system	Sous Chef Ltd	UK	Low-acid meat and vegetable ready-meal products	Coloreed bag-in-box filling into 2.5-l bags for distribution to the catering market Bosch form/fill/seal or filling into retail trays	Nominal 750 kg/h
	Nissei	Japan	Whole strawberries in syrup	Dai Nippon bag-in-box filling into 10-l bags for distribution to the high-quality catering and dairy market	Nominal 750 kg/h
	Confidential	UK	Fruit, meat and vegetable formulations	Confidential	Nominal 750 kg/h
	Confidential	Japan	Multi-product use		Nominal 750 kg/h

Table 4.5 Continuous-flow research systems

Customer	Location	Products	Throughput	Comments
R-Tech	USA	Multi-product use	Nominal 50 kg/h	To be installed in the Center for Ohmic Commercialization. Funded by consortium of food processing companies for developmental work and obtaining FDA/USDA approval
Confidential	Europe	Multi-product use	Nominal 50 kg/h	
Confidential	Japan	Multi-product use	Nominal 50 kg/h	

4.10 Conclusions

There can be no doubt that the ohmic heating system provides the food processor with the opportunity to produce new, high-added-value ambient-stable products with a quality that was previously unrealised with alternative sterilisation techniques.

The ability to heat particulates uniformly, without mechanical damage, combined with lower nutrient and vitamin losses and no fouling of heat transfer surfaces will ensure that ohmic heating will play a major role in the growing requirement for aseptic food products containing large particulates.

These advances in heat preservation technology have now been developed to the stage where systems are now in commercial operation throughout the world.

Acknowledgements

I am grateful to the Directors of APV Baker Ltd for permission to present this chapter and to the UK Electricity Council from whom the ohmic heater is licensed.

Reference

Brown, K. L., Ayres, C. A., Gaze, J. E. and Newman, M. E. (1984) *Food Microbiology*, **1**, 187–198.

5 The ERCA neutral aseptic system

C. J. ROBINSON

5.1 Introduction

The concept of form/fill/seal (FFS) technology has been one of the most significant growth areas in food packaging over the recent years. The most noticeable successes have been in the dairy and drinks industries but there has also been considerable expansion in other forms of food manufacture. The primary market for this technology was formerly in mainland Europe, especially France, but today FFS machines can be found throughout the world.

The original aim of FFS was to achieve cost savings. Firstly, materials supplied in reel form were significantly cheaper than preformed packaging. The manufacturing process was simpler and quicker, with reduced investment in converting machinery; transport and storage costs were also much lower. Secondly, a compact machine could achieve high outputs with minor operator involvement.

These features continue to offer significant manufacturing benefits; indeed, over the past decade market influences have added significantly to the importance of FFS. The convenience to the retailer and consumer of multipacks can be easily exploited without the penalty of additional over-wrapping and FFS also provides one of the lightest and most efficient form of packaging in today's food market. The chilled dairy and dessert cabinets in most European supermarkets are dominated by FFS multipacks, with certain product categories sold almost uniquely in this format. Ambient display areas also contain a wide selection of FFS long-life products including dairy, vegetables, fruit and juices, and snack products. The ERCA neutral aseptic system (NAS®) has been developed to produce a safe, economical and convenient package for both supermarkets and consumer, one that avoids the use of increasingly undesirable chemical sterilants for the disinfection process. Having achieved US FDA approval in 1982, the NAS System® has been widely adopted in all continents by some of the most prestigious food manufacturing organisations.

5.1.1 *Sterility without chemicals*

'Form/fill/seal' is an over-simplification of the process that on one machine thermoforms, labels, fills, heat-seals and cuts anything from one to over 40 containers for each operating cycle. The machines function on an indexing basis using a mechanical drive mechanism and hourly output is dictated by the indexing frequency, the size of forming area and the dimensions of the container. Once established, machine outputs remain constant and can range from 1000 to over 60 000 containers/h.

Figure 5.1 Typical range of ERCA FFS and NAS products.

The ERCA NAS System® has the unique advantage of operating with pre-sterilised materials, which are supplied in multi-layer, co-extruded construction; peeling away the top layer within the sterile machine environment exposes bacteria-free product-contact surfaces. ERCA not only eliminates the inconvenience, expense and danger of chemical sterilants but also reduces the risk of contamination of the product. Sterilisation of the materials is ensured by the high temperature/co-extrusion time during the manufacturing process. The sterility of the NAS machine is achieved by the use of dry heat—a minimum machine sterilisation by heat treatment is substantially greater than that required for spore destruction of the most heat-resistant microorganisms currently known. ERCA packages can be designed and produced in virtually any style, shape or colour, in sizes ranging from several grams to 2 kg. Since chemical sterilants are never used, the aseptic barrier packaging requirements can be met with a wide range of multilayer container and lidding materials. ERCA multipacks also achieve very high material utilisation because frame scrap between the containers is virtually eliminated.

The ERCA machine can also use a well-proven wrap-around labelling system that automatically bonds labels to containers as they are formed. This process adds side-wall strength to a container, as well as providing a very attractive and informative decoration. The wrap-around label can also provide a significant additional barrier property by reducing ultraviolet (UV) penetration of the container.

The ERCA NAS System® provides a non-tied packaging operation, leaving the user free to purchase the materials from several sources. ERCA has worked closely with several material suppliers to ensure that the highest levels of hygiene and good working practices are employed during the manufacturing process. The success of the NAS System® is dependent on the integrity of the materials. In general terms, the materials are relatively simple to make and can be easily sourced in any industrialised location.

5.2 Basic construction of the ERCA machine

5.2.1 Preparation of the materials

A major factor influencing the cost of FFS containers is the ability to purchase them in reel form; this reduces the converting process and the cost of the transport. The packer can also make more effective use of storage facilities, potentially reducing capital investment in buildings and so on. Once removed from their wrappings, the reels are mounted on wheeled dollies or simple mandrels, ready to be loaded into the machine. The packaging should be 'acclimatised' to the ambient conditions in the packing room. This is normally achieved by maintaining a 24 h stock in close proximity to the machine.

Table 5.1 NAS machine applications

Location	Range of equipment	Product range
Australia	NAS 1 NAS 3	Tomato paste
Canada	NAS 3	Desserts, soups, sauces
China	NAS 3 NAS 2 SRS	Tomato concentrate, fruit juice
France	NAS 2 NAS 3	Fruit cocktail, peaches, pears, milk-based desserts, cream, fruits, yogurt
Italy	NAS 3 NAS 4	Ice cream, drinks
Martinique	NAS 3	Fruit juice
Malaysia	NAS 3 SRS	Fruit juice
New Caledonia	NAS 2 SRS	Fruit juice, yogurt
New Zealand	NAS 3	Various products
Singapore	NAS 3	Fruit drink, desserts, soya milk
USA	NAS 3 NAS 4 NAS 1	Pudding baby food, barbecue sauce, apple sauce, soups

5.2.2 Machine operation

The plastic reel mounted on a wheeled dolly is precisely located at the entry to the machine. The sheet is partially unwound, looped around a balancing roller and, via side guides, is fed into the chain transport system. The machine is equipped with a pair of chains that run along the full length of the machine and return via a set of tensioning pinions. Each chain is fitted with a set of spring-loaded grippers, which are opened at the plastic in-feed and then allowed to reclose around the plastic, securing it firmly. At the discharge end of the machine, these grippers are again opened and the plastic waste is allowed to be drawn from the chains. In addition to simply moving the plastic, these chains, coupled with a precisely positioned support system, control the web to avoid distortion and to minimise the effects of expansion and contraction. This is especially important when the machine is producing parallel-sided containers; this control must avoid any sagging of the plastic across the web and also maintain a constant tension along the axis of the machine and along each edge of the material.

The transport chains must therefore always move an equal distance and ERCA has developed a magnetic pulley system that will compensate for any stretch or undue wear in either of the chains and ensure consistent movement on both sides of the machine.

The plastic is pulled into a heating box, which raises the temperature of the material to the point when it can be effectively thermoformed. Infrared heating elements are used to avoid contact with the material and to avoid the risk of cross-contamination. The heating box must generate an even distribution of heat not only across the plastic web but also through the material, and this is particularly relevant when thicker multilayer types or those with multiple colours are being used. The operation of the heating box is carried out in a series of heating zones, which will ensure this uniform temperature distribution. Heating control is of paramount importance not only to ensure correct forming but also to optimise material utilisation when starting or stopping the machine. The heating effect must be increased when the machine is started using cold plastic and similarly, when the machine is closed down, the heating box must be cooled to ensure that any plastic remaining in the box is not overheated, avoiding the risk of melt down or, more seriously, the risk of fire.

The heated web then indexes to the forming station. At this stage, the machine that produces multipacks may be fitted with an in-mould labelling device since it would be difficult—and in the case of wrap-around labels impossible—to apply them after the packs have left the machine. The reel of labelling material is placed on a wheeled dolly and located at the side of the machine. The material is wound over some tensioning rollers and is then slit into a series of bands corresponding to the number of cavities in the forming mould. These bands continue across further tensioning and balancing bobbins and are finally cut and coiled inside the mould cavities. Reel change-overs can be implemented without interrupting production by using an automatic splicing unit to bond together the old and new reels.

The labelling process uses a paper or laminate with a heat-sensitive adhesive on its inner surface and when it comes into contact with the hot plastic, the adhesive is activated and bonds the two materials.

One important function for the label is to attract attention and to display information such as bar codes, ingredient panels, and so on. In wrap-around form it also provides rigidity thereby minimising the thickness of plastic needed to form a container and it can also provide barrier protection. For example, in metallised form, the label is not only aesthetically pleasing but the aluminium layer provides a shield against ultraviolet light.

Moulding is a plug-assisted compressed-air forming process. The heated, softened plastic is pulled over a mould, which is usually machined to have several cavities and is maintained at a constant temperature by the circulation of stabilising fluid. The mould temperature is usually kept at around 15°C, although this can vary according to the material and the speed of the machine. An upper counter-mould clamps the plastic and the specially-shaped plugs of Teflon or similar material stretch the plastic into the cavities; the release of compressed air around the plugs completes the process by pushing the plastic into the final recesses of the mould. The hot plastic activates the label adhesive and then the cool walls of the mould reduce the plastic temperature so that it retains its shape.

Containers tend to be conical or parallel-sided designs, although ribs or engraving can be used to improve appearance or strength. In special cases, reverse-taper containers can be made, although this requires split-tooling to allow containers to be withdrawn from the mould.

The variety of products packaged on FFS machines requires a range of alternative dosing units; pasty products behave differently to liquids and fruit pieces, for example, need special consideration to preserve fruit identity and to avoid carry-over between containers and also to ensure accurate filling tolerances. Aseptic filling for long-life products has its own special technology to maintain bacteria-free operation.

The filling process is generally volumetric, although other applications include timing devices, flowmeters and filling by weight for dry products. It is standard practice on all ERCA machines that the fillers must be suitable for cleaning in place with no dismantling. Fillers are usually cleaned using a sequence of rinses with caustic soda and acid solutions and these can be followed by hot water, steam or chemical sterilisation treatments. Cleaning solutions are selected to suit the characteristics of the product; for example, nitric acid is often used for descaling milk-based deposits on dairy machines.

The filling station can be supplemented by a gas-flushing unit usually using nitrogen, which will displace oxygen during the filling process as an additional protection for sensitive products.

The filled pots then index on to the lidding station where a sheet of mono- or multi-layer material is applied over the filled containers. Supplied in reel form on a mandrel, this material can be aluminium, combinations of aluminium, paper and polyester or any one of a number of different polymers. The material is usually printed to enhance the appearance of the container and, in most cases, the design is matched to the container format.

In order to maintain synchronisation of the lidding design with the forming of the containers, and although today's printers can work to extremely tight tolerances, it is necessary to incorporate a means of adjusting the relationship between the two materials. This is achieved by printing the pitch of the lidding material design shorter than the indexing movement of the machine and then stretching the material into synchronisation. A typical comparison would be: machine (i.e. plastic) pitch 100.06 mm *versus* lidding pitch 99.89 ± 0.05 mm. The lidding print incorporates a registration mark that is detected by a photocell and the machine can compare the movements of the two materials and decide if stretching is necessary.

The lidding is stretched using a series of rollers that grip and extend the materials, and aluminium film is very easy to use because the stretching provides permanent extension. Multilayer materials have become more popular because they have better puncture resistance and 'peelability' because they incorporate one or more polymers in their construction. This also gives the materials an elasticity, which therefore requires precise adjustment of the machine itself and close monitoring of the materials to maintain the correct stretching process.

The prestretched material can then be date-coded using conventional rubber-stamp technology or increasingly popular ink-jet equipment. It can also receive some pretreatment of ultraviolet or infrared radiation as a form of bacteriological treatment. Reel change-overs can be carried out automatically using a splicing unit.

Polymer lidding can be thermoformed on the machine to provide an internal fitting shape to locate itself inside the container and providing a reclosable cover to protect the contents after opening.

At the manufacturing stage, the lidding web is pretreated with a heat-sensitive lacquer to match the characteristics of the plastic; the two materials are then bonded together in a press where a heated block provides compression and activates the lacquer. The heat seal is applied to the flange around the top of the container, which is supported from the underneath by a counter electrode. Precise control of the plastic is essential to guarantee an effective seal and to ensure that the movement of the lower part of the press, the counter electrode, does not crush the containers.

At this stage, the machine contains a web of filled, labelled and lidded containers, which must then be divided into the desired combinations. Cutting in singles or multipacks of up to six or more containers can be achieved, with the flexibility to vary the cutting patterns if required. Single containers are die-cut from the web leaving a lattice of scrap material, which is wound onto a reel for easy disposal. Multipacks generate less waste, using more of the available material to produce the containers. The typical multipack container offers the optimum material utilisation; its combination of surface area and plastic thickness allows an even plastic distribution around the walls and base of the container, which will provide the correct level of physical and barrier protection to the product.

Stainless steel is always used for those components that contact the product but other materials such as carbon steels, aluminium, polymers, and so on, are used in

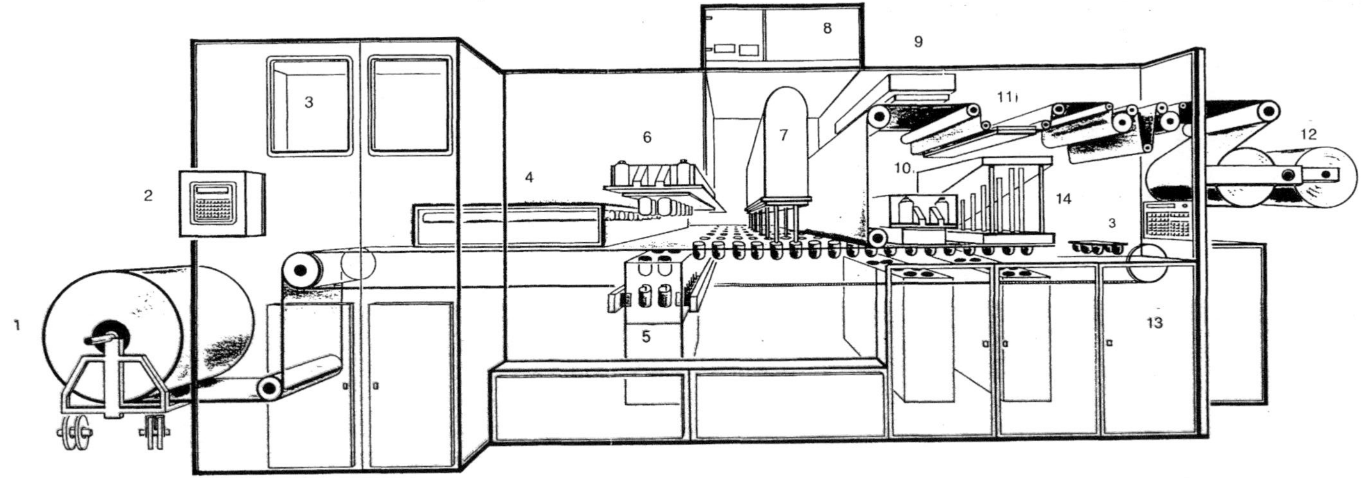

Figure 5.2 The standard ERCA hygienic machine. 1 = Plastic reel infeed; 2 = information printing; 3 = PLC; 4 = infrared plastic heating 5 = in-mould labelling; 6 = forming mould; 7 = volumetric filling; 8 = sterile air overpressure; 9 = infrared 'flash' lidding heat treatment; 10 = heat-sealing; 11 = date-coding; 12 = lid reel splicing; 13 = synchronised mechanical drive; 14 = cutting and trimming.

the general construction of the frame and drive mechanisms, where performance parameters such as weight, strength, heat transfer and so on have high priority.

5.3 Extension of basic machine to NAS® specification

The NAS® aseptic machine incorporates several additional refinements in order to provide a sterile packaging operation; firstly, the machine must be capable of being sterilised and thereafter must maintain that sterility during the production process. Furthermore, the machine must incorporate the facility to separate/delaminate the special presterilised materials.

The NAS® container and lidding materials are multilayer structures including at least two delaminable co-extruded layers. One of these layers would be delaminated film and the remainder of the material would be used to provide the container. The NAS® process strips away the delaminable films while maintaining the sterility of the surfaces of the materials that provide the internal product-contact areas. The delaminated films are pure materials that can be readily recycled by the supplier.

5.4 NAS® materials

The structures of the packaging materials are determined by several parameters:

1. The shape of the container.
2. The required shelf-life of the product.
3. The storage conditions (temperature, light, and so on).

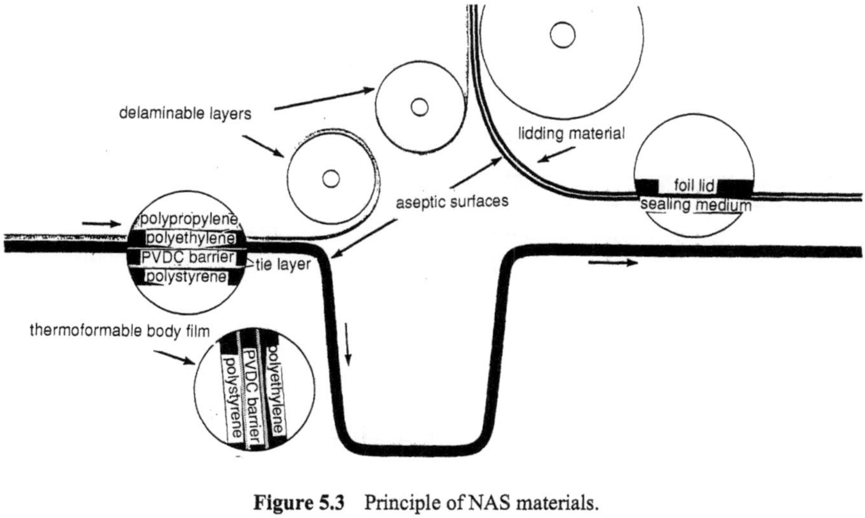

Figure 5.3 Principle of NAS materials.

4. The characteristics of the product (high/low acid, proteins, fats, and so on).
5. The transport conditions (any rough handling or transport over long distances).
6. Temperature extremes (frozen dessert applications or microwave heating).
7. Method of product consumption (e.g. straws, spoons).
8. Legislatory or environmental influences.
9. Cost.

5.4.1 Plastic base web

A typical extended-life package would consist of a polystyrene and polyethylene combination combined with a polypropylene delaminable layer. In contrast, a long-life container could be formed from a more complex structure involving polystyrene, PVDC and polyethylene, plus the polypropylene delaminable film.

The polymers that are used in the construction of the plastic are extruded separately and converge towards a feeder block and die. In this block the materials are proportioned in order to have the correct distribution of material in the various layers of the finished plastic, as defined by the intended application. It is thus after this point at the discharge from the die that film reaches its final thickness. This treatment makes it possible to raise the temperature of a material entering in the composition of the NAS® structure to 200°C for 5 min (from the feed point into the extruder down to the discharge of the die). Thus at every point of the interlayer between the delaminated film and the base web the respective surfaces have undergone a treatment ensuring their sterility.

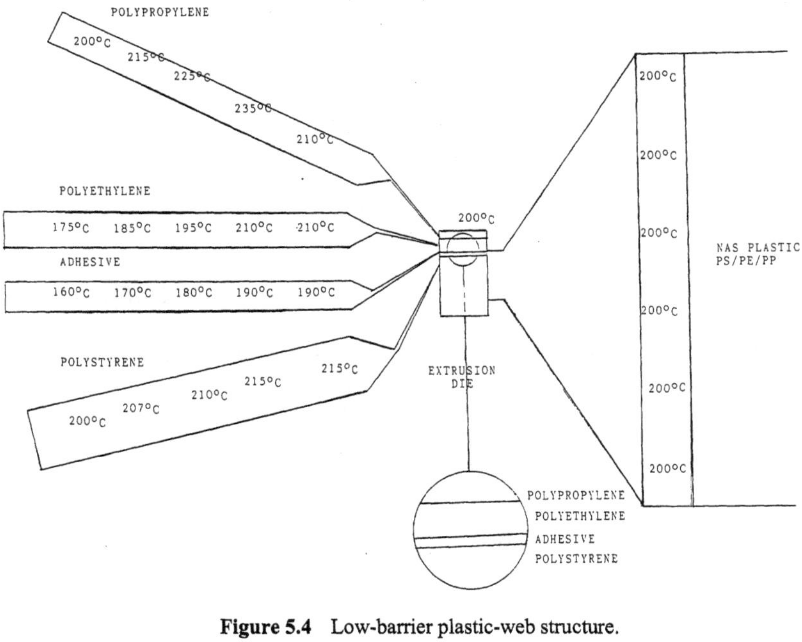

Figure 5.4 Low-barrier plastic-web structure.

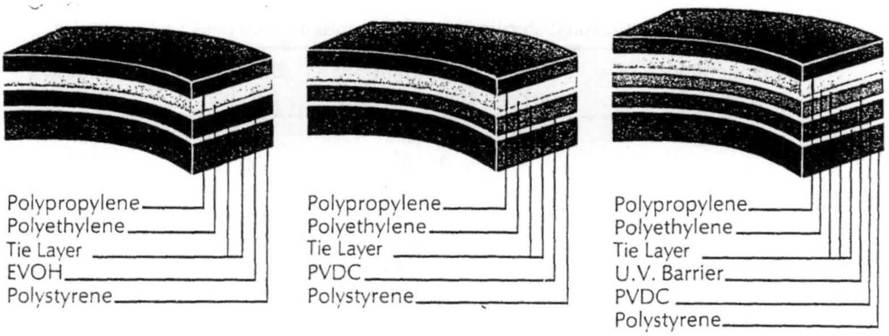

Figure 5.5 Plastic-barrier structure options.

The sterilisation table, established by the destruction curves for *Bacillus subtilis* var. *niger*, gives 5 s as the application time for a temperature of 200°C to obtain the destruction rate of 10^8.

5.4.2 *Lidding film*

The lidding film is manufactured in a similar process to the plastic whereby two delaminable polymers are co-extruded under conditions of time and temperature suitable to ensure sterility. They are then applied to a film of aluminium or aluminium/paper combination; depending on the co-extrusion equipment used the external delaminable layer may be either polypropylene or polyester.

5.4.3 *Verification of sterility of NAS® co-extruded materials*

The manufacturing process will ensure the sterility of the ERCA NAS® materials but in order to demonstrate the security of this system ERCA has carried out specific tests to confirm this point. A test facility was therefore established to reproduce the extrusion conditions (in other words to apply the recognised temperature/time conditions).

Samples of NAS® materials were prepared by delaminating the peelable layer and placing on the normal product-contact surface temperature probes and also samples of *Bacillus subtilis* (var. *niger*) with a concentration of 2×10^8. The peelable layer was then replaced and the contaminated samples were pressed in between two thermostatically controlled plates so that the temperature of the product contact surface was raised to 200°C. This temperature was maintained for

Table 5.2 Destruction by dry air of *Bacillus subtilis* var. *niger*: time of destruction of 10^n spores

Temperature (°C)	(°F)	D (s)	Time of destruction of 10^n spores ($n \times D$) $n = 2$	$n = 3$	$n = 4$	$n = 5$	$n = 6$	$n = 7$	$n = 8$
140	285	141	4 min 04 s	7 min 04 s	9 min 24 s	11 min 45 s	14 min 06 s	16 min 27 s	18 min 48 s
150	300	64	2 min 08 s	3 min 12 s	4 min 16 s	5 min 20 s	6 min 24 s	7 min 28 s	8 min 32 s
160	320	16	32 s	48 s	1 min 04 s	1 min 20 s	1 min 36 s	1 min 52 s	2 min 08 s
180	355	5.7	12 s	17 s	23 s	28 s	34 s	40 s	46 s
190	375	1.9	3.8 s	5.7 s	7.6 s	9.5 s	11 s	13 s	15 s
200	390	0.63	1.3 s	1.9	2.5 s	3.1 s	3.8 s	4.4 s	5 s

Table 5.3 Test results: sterilisation of precontaminated packaging samples

Temperature reached (°C)	Time (min)	Initital contamination*	Bacterial survival	Destruction rate
199°C	5	2×10^8	0	10^8
		2×10^8	0	10^8
		2×10^8	2	10^8
		2×10^8	3	10^8
		2×10^8	<1000	10^8
199°C	7	2×10^8	0	10^8
		2×10^8	1	10^8
199°C	10	2×10^8	0	10^8
		2×10^8	0	10^8
		2×10^8	0	10^8
		2×10^8	0	10^8
		2×10^8	0	10^8
		2×10^8	0	10^8
		2×10^8	0	10^8
		2×10^8	2	10^8
204°C	5	2×10^8	0	10^8
		2×10^8	0	10^8
		2×10^8	0	10^8
		2×10^8	1	10^8
		2×10^8	0	10^8
204°C	7	2×10^8	0	10^8

* With *Bacillus subtilis* var. *niger*

a maximum of 5 min representing the passage of time of the material in the extruder and through the extrusion die.

Examination of a sample after this treatment process confirmed that the destruction rate achieved in the co-extrusion conditions is at least 10^8 for *Bacillus subtilis* (corresponding to 10^9 for *Clostridium sporogens* PA3679 or 10^{13} for *Clostridium botulinum*).

5.5 NAS® machines

5.5.1 General description

The sterile NAS® films are delaminated aseptically, before entering the sterile tunnel. The plastics film is thermoformed and the container so produced is filled before the cover film is applied and given a hermetic seal. These latter operations are carried out in sterile conditions. Like the films during their manufacture, the machine is sterilised by dry heat applied by built-in heating elements at certain points in the machine (e.g. filler, forming, and so on) and by the injection of air superheated to 250°C in the following areas: forming, filler and tunnel.

The destruction curve for *Bacillus subtilis* and the decimal reduction figures for the main bacteria make it possible to calculate the sterilisation parameters. The sterilisation of the machine is ensured by applying a temperature of 160°C (320°F) to the coldest point for 15 min. The respective sterilisations of the sterile areas will

Table 5.4 NAS® machine sterilisation parameters

Microorganism	Temperature (°C/°F)	Application time (s)	Sterilisation effectiveness
Stearothermophilus 1518	160/320	110	6
Polymyxa PSO	160/320	80	6
Clostridium sporogens 3679	160/320	25	6
Clostridium botulinum 34B	160/320	23	6

be examined in the detailed descriptions of the tunnel, the forming station and the filler. After the rise and the maintenance of the temperature, the parts taken up to 160°C are cooled by circulation of cold sterile air.

At the end of the sterilisation cycle (temperature rise – temperature holding – cooling) the sterility of the machine is ensured by constant pressurisation with sterile air. The compressed air supplying the filler and the forming station, as well as the air delivered by a turbine fan are presterilised by passing through a set of filters (de-oiling, retention of solid particles down to 0.01 μm, and elimination of odours by an activated carbon filter). A safety device checks the continuous supply of sterile air and controls the shut-down of production in the event of inadequate supply.

5.5.2 Plastic film introduction

The sterile ERCA NAS® machine is based on the use of sterile delaminable materials. In order to maintain this sterility, the opening of the films to expose the sterile surfaces must therefore be carried out aseptically.

The co-extruded film enters the tunnel and the delaminable polypropylene film is drawn away following an angle given by a delamination blade. It is rolled up by a rotary mandrel, simplifying its recovery and later recycling. Simultaneously, the thermoformable part, guided by the chains, moves on to the heating chamber. The two sterile surfaces freed by the delamination form a boundary to the sterile area, the maintenance of the sterility of the latter being ensured by the introduction of the sterile air pressurisation.

The operating tunnel is formed on the top part by the various tools and equipment (forming, filling and so on) and by close-fitting covers, the bottom part of the tunnel being provided by the plastic web. The elevated temperature required for sterilisation is too high for the plastic so, during this phase, a sheet of 80 μm aluminium foil is introduced into the machine. At the start of production, therefore, the first reel of plastic must be aseptically bonded to the aluminium.

A critical feature of maintaining sterility is to control the introduction of the first reel of plastic web and thereafter the change-over of reels, to ensure that there is no introduction of bacteria on the materials themselves. The polypropylene film is delaminated for a length sufficient to be wound onto its recovery mandrel. The partially exposed sections of aluminium and plastic, which may have become contaminated during the joining process, are then indexed into a flash box containing a bank of infrared elements, which provides a sterilising effectiveness of 10^6.

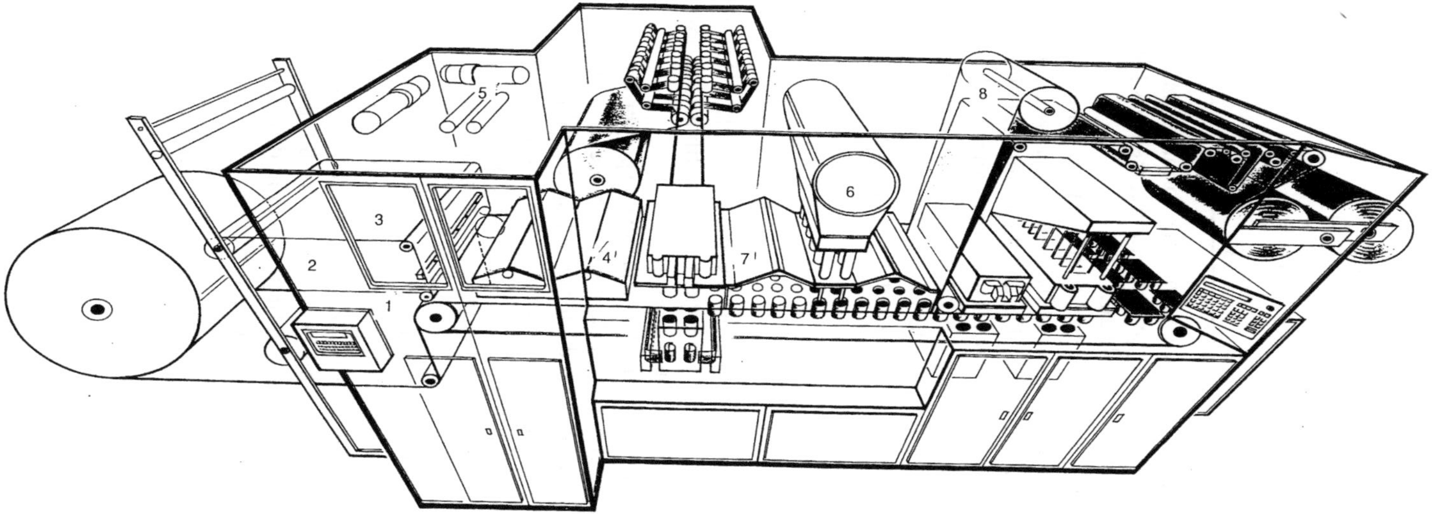

Figure 5.6 The NAS machine. 1 = Multilayer plastic infeed; 2 = PLC control; 3 = plastic cover film delamination; 4 = 'flash' and plastic heating; 5 = sterile air preparation; 6 = aseptic filling; 7 = aseptic tunnel; 8 = lidding cover film delamination.

A Teflon sheet protects the polypropylene film from the effects of the infrared heating and also ensures an airtight seal at this delamination point.

The production time for a reel of plastic may vary, according to the size of a reel and the throughput of the machine; in general, short reels are discouraged to minimise changeovers, which result in lost manufacturing time and waste of materials. Sterile production runs can in some cases last for more than 24 h and reel changes are a routine operator function. The reel already in use and the reel to be introduced are cut at right angles on a special splicing table. The two ends so produced are placed face to face; the leading edge of the replacement reel is delaminated over 5 mm. The ends put thus edge-to-edge are fastened on both faces—the upper polypropylene outer surfaces and the lower polystyrene outer surfaces—by means of a self-adhesive aluminium-based heat-resistant tape. After delamination of the polypropylene film, which is partially exposed to contamination during the joining operation, the sections of the two reels are indexed under the bank of infrared elements for resterilisation.

5.5.3 Lidding film introduction

The material combination intended to supply the sterile cover is separated into two films. The delaminable film turns round the lower angle of a delamination blade, then goes on to a rotary mandrel on which it is rolled ready to be recycled, and the aluminium/polyethylene combination is drawn towards the sealing station, where its internal surface will form the sterile face of the cover of the packing so produced.

During the sterilisation of the machine, the delamination of the cover film has already been achieved whereby the polypropylene film is already rolled on the rotary mandrel, and the aluminium/polyethylene film has been introduced so far as its outlet from the sealing station. A protective sheet of Teflon prevents the destruction of the thin film from the temperatures maintained during sterilisation.

Lidding-reel changes take place less frequently than plastic change-overs; nevertheless, bacteria-free changes must still be made. The principles used are similar to those employed at the plastic splicing unit where the two sets of materials, delaminable film and aluminium/polyethylene, are taped together and then subjected to infrared resterilisation.

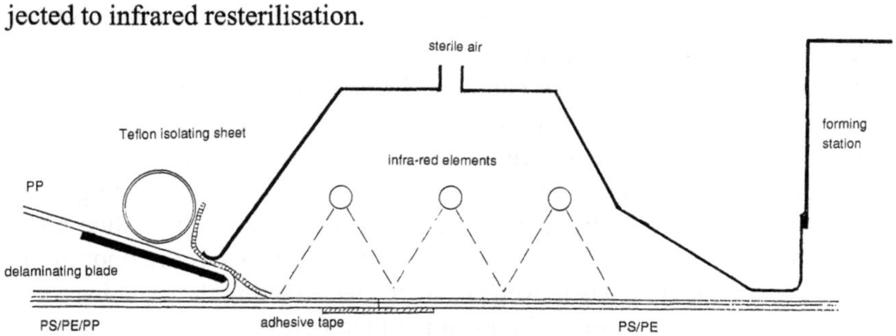

Figure 5.7 Plastic film delamination.

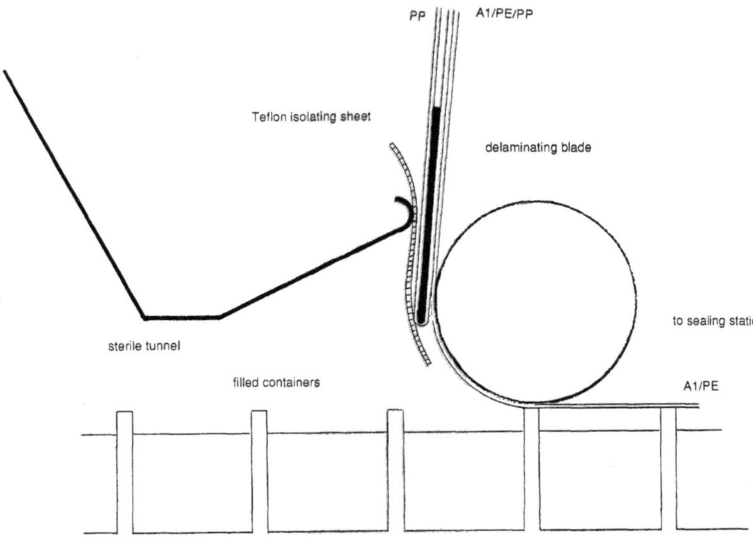

Figure 5.8 Lidding film delamination.

The metallised self-adhesive tape used in the bonding of both base web and lidding materials must be eliminated from the saleable product; in the case of the plastic, it would not survive the thermoforming process. The machine therefore features detection devices, which, allied to the PLC controller, interrupt the various functions of the machine to ensure that the splices are moved through the machine without touching the product or any product-contact surfaces, and to ensure that the operator is instructed to remove the materials at the discharge of the machine.

5.5.4 Sterile tunnel

The tunnel is limited to the sterile filling area, that is from the inlet of the heating box to the inlet of the sealing station. It ensures the maintenance of the sterility of: (i) the thermoformed containers before filling; (ii) the product at the filler; and (iii) filled containers until their closure by the sterile cover film.

In the widest sense, the tunnel could be defined as that part of the machine maintaining the sterility of the materials from their delamination until the closure of the filled containers. It therefore includes the plastic flash box, the plastic heating chamber, the forming station, the filler and the free areas dividing these items. It has been stated earlier that, during sterilisation, a sheet of aluminium foil is pulled through the machine in the chain grippers, to act as the floor of the tunnel; in production, the plastic web provides this, enabling the machine to be kept under a positive pressure and preventing this ingress of airborne contamination.

The boundaries of the tunnel are:

1. The aluminium or plastic sheet.
2. The operating components of the machine.
3. Upper covers, which are removable for maintenance and so on.
4. The delaminating knifes/Teflon films.

The sterile overpressured air enters through the tunnel covers and is allowed to escape past the Teflon film and the transport-chain grippers. The distribution of the air must be carefully controlled to ensure that there are no low-pressure zones in the tunnel and also to provide rapid heating and cooling during and after sterilisation processes.

The various parts of the tunnel are sterilised by dry heat, according to the sterilisation tables previously discussed (see Figure 5.6). The rise in temperature and the retention at the prescribed level are provided for by the circulation of superheated air at 250°C. Internal resistors in the lower part of the filler, placed below the filler hood, supply additional heating for this block that the superheated air alone cannot bring up to a temperature of 160°C. The machine is maintained at this temperature for a period of 15 min.

At the start of sterilisation, the lower part of the tunnel is closed by introducing the aluminium foil, long enough to cover the area starting from the plastic film delamination point so far as the sealing station outlet. The operation of the plastic flash box sterilises the part of the tunnel between the delamination point and the inlet to the plastics heating chamber. The temperatures reached during sterilisation are checked by four probes (forming station, filler block, plastics heating chamber, and the end of the tunnel at the sealing station), while three probes provide for the regulation of the superheated air temperature (supplying the tunnel, the filler and the forming station). The temperature rise continues until the coldest point has reached 160°C, and it is maintained for 15 min. The cooling is carried out by a supply of cold sterile air for 15 min, which supplements the circulation of cooled water inside the filler. At the end of the sterilisation, the start-up of the transfer chains removes the aluminium foil closing the tunnel to be replaced by the thermoformable part of the sterile NAS® film.

The tunnel can be maintained under an overpressure of a gas such as nitrogen in order to eliminate oxygen contact with the product. It also ensures that the headspace in the container, above the product, is oxygen-free—particularly relevant for certain types of product.

5.5.5 Forming

The forming of the containers is achieved by the same methods as in non-sterile machines, except that the air supply is constructed to respect the condition of sterility. A filter assembly supplies sterile air permanently during the operation of the machine and the forming air is supplied from an individual filter unit. The forming air is superheated to a minimum of 300°C before entering the forming station. Similarly, the air coming from the sterile filter placed in the upper part of the forming station is superheated in such a way as to pass through the filter element at 250°C.

The temperature rise is accelerated by the incorporation of heating elements inside the heavier components, and after maintaining the sterilisation temperature of 160°C for a period of 15 min, the circulation of cold sterile air provides cooling of the forming assembly.

5.5.6 Filler

A table identical to that for the tunnel is applied to the sterilisation of the filler, to obtain a sterilisation effectiveness of 10^6 by the maintenance of a temperature of 160°C for 15 min. The necessary addition of heat to increase the temperature of the parts making up the filler is supplied by superheated air into the upper part, the lower part incorporating heating elements in the side walls. At the end of the sterilisation holding time, sterile cold air is circulated to cool the filler, which can be completed by circulation of sterile water in the phase preceding the filling.

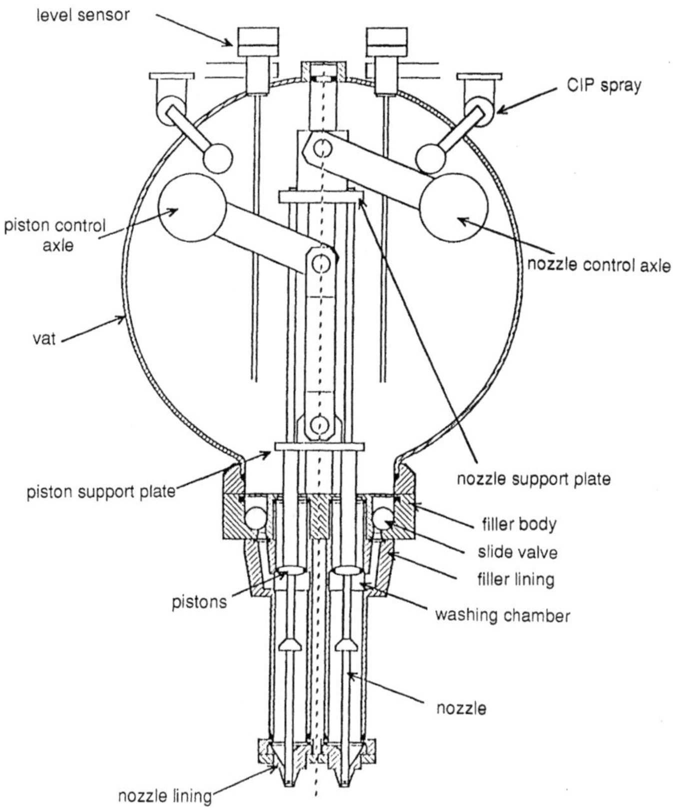

Figure 5.9 Sterile filler for liquids.

A range of fillers is available to suit a variety of different products. Construction is the monoblock type for the following reasons:

1. Reduction in the number of product connections: one inlet is required, controlled by a multiport valve, which permits the cleaning and sterilisation of the filler and the product supply pipework.
2. Reduction in the number of operating mechanisms: only two mechanisms are required for a multi-head filler, minimising the risk of contamination across seals or a rotating or reciprocating shaft.
3. Control of sterilisation: the process of monitoring sterilisation temperatures is more secure on a monoblock unit when compared with several individual modules.

The product inlet is located in the side of the filler body to reduce splashing and avoid foaming. The operating mechanisms feature sterile air overpressured seals to maintain an aseptic barrier against outside contamination.

The sterile air overpressure in the filler bowl, in the process of maintaining sterility, also assists the product flow through the filler; in order to ensure consistent filled quantities—usually measured by weight—the static pressure in the filler is accurately controlled using a modulating PLC-controlled air regulator featuring digital display with preset functions with alarms. A lack of overpressure will result in lost sterility; the system therefore monitors the value of the overpressure and stops the machine (also preventing restart) if this value falls below a minimum level.

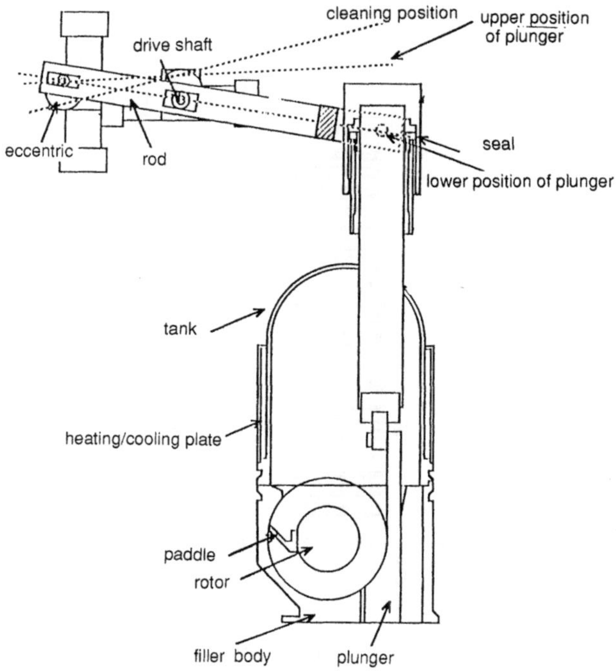

Figure 5.10 Sterile filler for particulate products.

The pallet filler has been especially designed for dosing particulate products. These particulates must be in the form of a pumpable product; in other words the pieces must be carried in a dressing or sauce. The combined solids/liquid is fed to the body of the filler to be distributed into the thermoformed containers. The filling principle is volumetric and it is important to ensure that the particulates do not settle at the bottom of the filler. A series of rotating pallets or paddles continuously sweeps the bottom of the filler delivering the particulates with the liquid into the dosing chambers. At a particular point in their rotation the pallets slice through any large pieces that may be overhanging the chambers, thereby maintaining consistent volumetric dosing. This minimal cutting operation also avoids the risk of large pieces remaining in the filling chamber, potentially dripping onto the heat-seal surfaces of the containers and ultimately resulting in a product failure. The filler will accept particulates of 25 mm^3, depending on their exact shape and hardness.

5.6 Product preparation

ANAS® machine is connected to the aseptic processing plant via a sterile inlet valve to the filler. In most cases, product is supplied from a continuous process from any of a number of equipment manufacturers. In certain cases, however, food

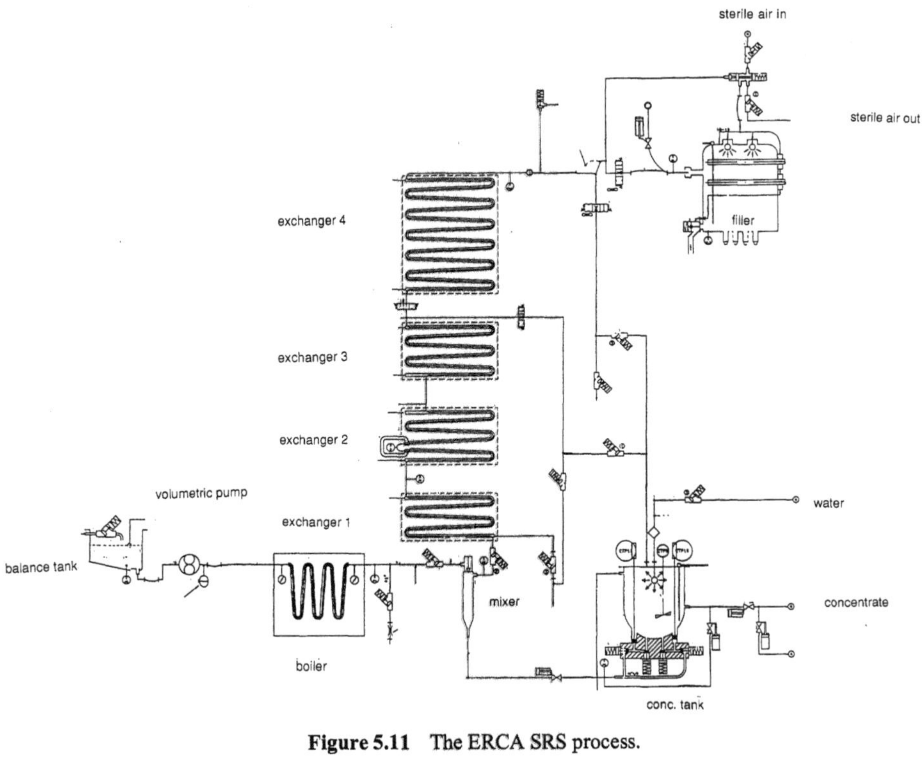

Figure 5.11 The ERCA SRS process.

manufacturers prefer to work with one supplier for the complete range of equipment— especially those in remote areas who do not have resources or access to modern technologies. In 1990, for example, ERCA supplied to China two lines for the production and packaging of almond-milk drinks and fruit juices. The equipment starts with grinding units to pulp apricot kernels, progressing to mix preparation using UHT processing on an SRS unit and finally packaging on an NAS machine; in addition, chilled water and compressed air assemblies were supplied to provide a complete installation requiring two simple connections for potable water and electric power.

The SRS®, or sterilising regeneration system, is an independent sterilisation unit for food products in which the dry extract is blended at a maximum rate of 25% (mainly drinks). The SRS® principle is a direct sterilisation process based on the mixture, at high pressure, of a concentrate (concentrated base or powder) with superheated water, which brings the necessary heat to the product for instantaneous sterilisation. Its operation is based on the following principles: (i) production of super-heated water; (ii) blending; (iii) UHT process; (iv) cooling; (v) cleaning (CIP); and (vi) process control.

5.6.1 Production of superheated water

This consists of a boiler whose capacity is adapted to the flow of the unit, and which is capable of delivering a sufficient quantity of heat to provide a continuous flow of superheated water and to make the whole unit completely independent for the sterilisation, production and cleaning phases.

5.6.2 Blending

This is a mixture of superheated water with the concentrate, comprising:

1. A volumetric pump for overpressure water capable of supplying the water at constant flow and pressure. Since the SRS® unit works on the principle of overheated water, the backflow of this pump is always superior to the vapour pressure of the water heated at the temperature of the boiler unit.
2. A volumetric pump for the concentrate, operating at high pressure to guarantee the flow of the concentrate in such a way as to obtain the exact mixture required for the final product.
3. A mixer under turbulent regime, and at high pressure, for the concentrate and the superheated water (constant pressure is maintained to avoid evaporation).

5.6.3 UHT process

In this process, permanent regulation of the sterilisation temperature and of the temperature of the sterile product at output is required. The SRS® is designed to provide variable sterilisation temperature and holding time according to the products. It can handle different products such as dairy products, fruit juices, soups, puddings, and so on, whose sterilisation norms require different values; these requirements are achieved at computer level, by simply calling the appropriate programme (depending on the product required).

5.6.4 Cooling

The sterile product at the exit of the holding phase is quickly cooled by parallel flow, in one or two exchanger tubes with cold water, followed by contraflow in one exchanger tube with cool water for the final cooling of the product. This indirect cooling provides a brutal thermal shock at the exit of the heat treatment, preserving the quality of the product, maximises heat recovery during the final cooling and enables fine adjustment of the temperature at the exit of the plant.

A good thermal balance of the installation is ensured by the recuperation and recycling of the cooling.

5.6.5 Cleaning (CIP)

The cleaning arrangement takes into account the total circuit between the tank starting the SRS® and the exit of the filler of the ERCA NAS® machine.

5.6.6 Process control

The SRS® unit is controlled by a microcomputer through a powerful PLC. No product recycling takes place, whatever the production stage. This management unit uses programmes to control all the working phases, regulation of the process according to the product characteristics, cleaning and sterilisation and defect analysis.

5.7 Future developments

FFS technology will continue to offer the existing benefits of low-cost manufacturing and packaging and the more recent developments in UHT processing—particularly those suitable for products with large particulates—will encourage further adoption of the NAS® aseptic packaging technology. Environmental issues will have a significant impact on future developments: already FFS containers use the minimum amount of materials and the production of multipacks need not involve additional overpackaging, but it will be necessary to progress further in the adoption of recyclable materials. Short-life, chilled, dairy multi-packs are manufactured in PS (polystyrene) plastic and developments have now reached a stage where lidding and labelling are now available in the compatible materials, providing the opportunity for industrial recycling of any waste from a machine or domestic recycling by the consumer. Multilayer materials are more difficult to recycle and ERCA is constantly working with material manufacturers to identify new combinations of materials that will satisfy future ecological standards. Future machine developments will include PLC extensions to improve bacteriological security, operating efficiency and simplicity, and PLC extension for improved management information and control.

Filler developments will enable ERCA to exploit new business opportunities—particularly for convenience applications such as snack foods, drinks and portion-controlled products and in other areas of the grocery trade.

Operator and technician training, supported by the relevant documentation, is essential for the successful operation of this equipment and a comprehensive after-sales service, providing technicians and spare parts, will ensure high levels of machine availability and efficiency.

Figure 5.12 The ERCA NAS machine.

6 Microwave processing and package integration

R. E. COLES

6.1 The technology revolution

The possibility of using microwave technology for food processing is hardly new. Some of the first applications were suggested soon after World War II had ended, during which the development of the magnetron continued and its subsequent use in radar occurred. World War II sailors soon found that the warmest place 'aloft' was close to the rotating radar scanner on board naval vessels (Freedman, 1985). Later, the idea of a 'box' around this thermal energy source was to follow. Twenty years passed before, in the mid-1960s, industrial microwave pasteurisation began to be used for bakery products. Microwave sterilisation studies also began at this time (Decareau, 1991) with independent work taking place simultaneously in Europe and the USA. Some of this early work has moved into continuing commercial production but not just for the food industry. Even today, the major uses of high-frequency energy are for different materials, all of which are in other (non-food) industries such as thermoplastics sealing, rubber vulcanisation, ceramic tile processing and so on.

Now such a technology revolution is beginning to offer real opportunities for today's food industry. The advent of new processing techniques and packaging technologies is beginning to create considerable interest. Microwave processing has the potential to be among the foremost of these advances, bringing with it packaging opportunities. Prepackaged foods are beginning to appear—either pasteurised or, more recently, sterilised using microwaves. With the possible exception of ohmic heating, microwave sterilisation avoids one major problem in the aseptic process in that there is no need to pump (and damage in the process) large particulate-containing food products through the system. The world's first two continuous-flow, pressurisable microwave systems for prepackaged foods are both operating commercially in Europe.

Some superior quality microwave-pasteurised, fresh pasta products have begun to be commercially produced (Figure 6.1). However, while it is known that research and development work indicates that microwave sterilised products can match their frozen (counterpart) controls (Decareau, 1991), the quality of the few products commercially available in the marketplace today leaves much to be desired. Yet, in theory, the products should be of considerably higher quality than is actually being achieved.

Decareau (1991) comments that 'the microwave process differs from conventional thermal processing in that the come-up time (CUT) to process temperature should be shorter'. He adds 'clearly some important element of the process has either been neglected or misused'.

Figure 6.1 Superior quality ready meals—pasteurised using the Berstorff continuous microwave processing system.

Another key factor in the successful microwave processing of prepackaged foods is the choice of packaging. This is considerably less well heralded. Yet the need for correct choice of packaging material and container geometry is equally vital for a successful outcome. The wrong choice has often been behind consumers' problems encountered during microwave product reheating. This same incorrect choice can represent the difference between success and failure when using microwave processing.

6.2 Consumer influence

New and exciting types of microwavable products are rapidly emerging, each having its own unique characteristics and properties, in an attempt to capture consumers' interest and appetite. Lifestyles have changed. Consumers today are more health-conscious and are seeking foods that are safe, of high quality, microwavable, fresh-tasting and yet are free of both additives and preservatives. The key factors among the growing wave of these new product entries is a focus on the maintenance of high quality, the extension of shelf-life, and convenience.

The extension of shelf-life is seen as benefiting the manufacturers, retailers and consumers alike. The portfolio of convenience products that have been recently developed includes refrigerated gourmet meals as well as salads and desserts. All of these are intended to meet or exceed growing consumer needs and desires. Many of these have been manufactured using processes other than by microwave but serve to indicate the tremendous range available. Some may never prove suitable for microwave processing. However, most present a challenge that manufacturers of microwave systems, packaging and food can meet collectively.

The microwave revolution is one of the fastest, if not the fastest, growing area of consumer interest. This trend is providing the impetus for both large and small manufacturers alike to create safe, yet innovative, microwave packaging and processing technology.

6.3 Packaging needs

Microwave processing systems require the packaging materials chosen to maintain control of the micro-environment present in the container. This is in addition to their normal function of protecting the product until opened for use by the consumer. Hence the packaging systems need to be designed to provide protection against environmental hazards (e.g. oxidation, moisture gain or loss, fat loss and so on). They also need to be reliably sealed to withstand the stresses that build up during processing, as well as those encountered throughout the distribution chain.

The packaging materials suitable for use with microwave thermal stabilisation systems can range from flexible packaging co-extrusions and/or laminates to premade, semi-rigid containers. The latter, after filling, are usually sealed with a flexible packaging membrane. For functional purposes, in addition to providing physical protection, sealed containers are then covered with a lid or packaged inside a paperboard retail carton or sleeve. In theory, a single sealed membrane should offer an adequate degree of protection if it has the same composition as that used for the container. However, care must be exercised in devising the package opening system. Any need to facilitate opening by the presence of 'tear-line' type opening devices needs to be carefully examined. Only with the very recent arrival of laser-controlled 'half-depth' cuts (Coles, 1991) can manufacturers consider this option; food safety considerations are paramount—especially for shelf-stable products.

Today's packaging industry is working hard to meet the challenges posed by developing systems to meet the needs of high-temperature/short-time (HTST) processing. These are fairly well-established for products such as high- and low-acid dessert and savoury fluid products. The lessons learnt by both packaging-material and filling-equipment producers are now being adapted to handle HTST, microwave-processed, products. Highly sophisticated non-destructive on-line testing methods are needed to detect flaws that would otherwise affect the integrity of the package. This is no easy demand. Scanning the sealed package surface is fraught with problems. Where aluminium foil is part of the closure system, the risk of an undetected pinhole-sized leak is doubly difficult to spot.

Examining the package seal areas before sealing can help but is not infallible. Water, for instance, can remain undetected by optical scanning systems. Vacuum-pressure tests on sealed containers are another option in common practice. Here, the nightmare could be the weakly sealed package that just passes an on-line testing regime but then fails during the subsequent handling pattern.

6.4 Packaging materials for microwaveable foods

These are one of three types: (i) microwave-passive; (ii) microwave-active; or (iii) microwave-reflective.

Until now, it is microwave-passive materials that have been singularly used to contain foods for industrial microwave pasteurisation or sterilisation. Microwave-active packaging materials absorb energy and heat directly in a microwave field. They are known as 'lossy' dielectrics—the 'lossier' they are, the better they absorb. Hence their rapid introduction for use in domestic microwave cooking or product reheating. Similarly, microwave-reflective materials, such as metals, provide the debate on their effectiveness. At present, their use too has been confined to microwave domestic cooking or reheating applications. Future use for both microwave-active and -reflective materials could well find new applications. They have the potential to provide differential heating—when carefully controlled microwave processing of multi-component foods is required.

6.4.1 Microwave-passive materials

In addition to the accepted consumer-oriented functions of a package (i.e. contain, protect, inform, sell, provide consumer friendliness, etc.), George *et al.* (1991) state that 'microwavable food packaging should:

1. Allow uniform heating throughout the product.
2. Withstand the temperatures for microwave only and combination oven use.
3. Be safe and simple to both handle and open (especially when hot).
4. Not impart any taint during cooking.
5. Meet legislative standards for the migration of package components into the food.'

Although this list was compiled for a UK government-funded research programme on heating in domestic microwave ovens, it applies equally to packaging materials used for industrial microwave processing.

Microwave-passive materials include ceramics, glass, paper and plastics. All of these are transparent to microwaves. They do not absorb heat directly, or only very little compared to heat absorbed by foods (Table 6.1). Thus packaging materials used to make containers for microwave processing are in this category since they allow the food to absorb the maximum amount of microwave energy available.

Table 6.1 Absorptive properties of common foods and microwaveable materials[1]

Microwaveable material	Percentage absorption	Microwaveable material	Percentage absorption
8% Saltwater	1.3	Soda lime glass	0.02
2% Saltwater	0.9	Thermoplastics/thermoset plastics	0.01–0.02
Vegetables	0.25–0.35		
Meat and fish	0.25–0.35	Borosilicate glass	0.001
Distilled water	0.12–0.15	Alumina	0.0006
Oils and fats	0.06		

[1]Source: G. Andrews, Gamma Consultants, Bromley, UK. MW PACK 88(PIRA).

Table 6.2 Common microwave-passive packaging materials

Material	Temperature resistance (°C)
Ceramics	> 500
Glass	> 500
Crystallised PET	220–260
PET-coated board	210
Polypropylene	160
High-density polyethylene	130

The commonly available materials used in microwave packaging applications are shown in Table 6.2. One key requirement of those materials used during pressurised, industrial microwave sterilisation is to resist the temperatures encountered; these exceed 100°C. This is thus likely to be more demanding than for domestic microwave cooking or reheating applications where the temperature in high-water-content foodstuffs will rarely exceed 100°C.

In practice, for reasons of both cost and required package performance, most packaging materials chosen for extended-life foods are multilayered plastics structures. Here, the benefits of several different polymers can be varied to meet specific needs. These range from providing requisite gas and moisture barrier for shelf-life purposes, through container manufacturing and filling-line performance, to adequate thermal properties required for both industrial microwave processing and domestic microwave cooking or reheating. They usually consist of three or more layers, for example an oxygen-barrier polymer, a structural or so-called 'carrier'

Table 6.3 Semi-rigid packaging materials for in-package microwave processing. (Source: Lawson Mardon Flexible, Midsomer Norton, UK.)

Material	Food type/application
PP/EVOH/PP	Sauces, ready meals and pasta
PP/PVdC/PP	Sauces, ready meals and pasta
CPET/APET	Sauces, ready meals and pasta
(HT)PS/EVOH/PP[1]	Sauces, dips
PP/filled PP/EVOH/ filled PP/PP	Ready meals developed for use on thermoform/fill/seal lines
PS/PVdC/PS	Dip sauces, pasteurised cheese
PS-PPO/EVOH/PP[2]	Sauces, ready meals and pasta

[1] Heat resistant polystyrene.
[2] Ultra-high heat-resistant structure.

polymer plus a tie layer that bonds the barrier layer to the structural layer. Barrier polymers are expensive and, alone, do not usually have the necessary mechanical properties such as tensile and impact strength, puncture resistance, high moisture barrier or adequate sealing characteristics. Table 6.3 includes both the commonly used combinations in use to date for containers plus some of the very latest additions. Again, all are microwave-passive materials. Typical membrane-lidding structures for use with these containers are shown in Table 6.4.

The structures listed in Table 6.4 are not the only ones in use. Others include:

1. PET/PP/EVOH/PP
2. Camclear O/PP—where Camclear O is an AlO_x-coated PET with an O_2 permeation of 1–3 $cm^3/m^2/24$ h at 23°C and 50% relative humidity and an H_2O barrier of 1–4 $g/m^2/24$ h at 38°C and 90% relative humidity. Camclear and Camcoat are trademarks of Camvac Ltd, Thetford, UK
3. Camcoat OPP/PE
4. Camclear M/PE—Camclear M is an improved moisture-barrier variant.
5. Met. PET/PP
6. PET/PVdC/PE
7. PET/SiO_x/PE

Multilayered, flexible packaging materials are also suitable for industrial microwave processing—either as the membrane lidding for a semi-rigid container (Table 6.3) or, alone, in the form of a shallow, profiled pouch (Table 6.5). This latter style is popular in Japan, either as a four-fin sealed pouch or bottom-gussetted, stand-up pouch. Changes are being pursued such as the replacement of OPP/aluminium foil/PE by silicon-oxide-coated PET-containing laminates such as PET/SiO_x/PE.

In Europe, some of the first products to be microwave-processed used flexible packaging on in-line type, thermoform/fill/seal machinery. Whether as lidding, pouch, or thermoforming, material combinations will largely exclude aluminium foil. Only microwave-passive materials are chosen. In a sealed pouch, these are the only options that will allow microwaves to penetrate the product. Where aluminium foil-containing structures are present as the sealed membrane lidding on a microwave-

Table 6.4 Flexible packaging materials for membrane lidding (oxygen permeation in $cm^3/m^2/24$ h at 23°C at about 65% relative humidity. All samples were retorted and dried before measurements taken. (Source: Lawson Mardon Flexible, Midsomer Norton, UK.)

Structure	Dry	30-min contact with boiling water	45-min contact with steam at 121°C
PET/aluminium foil/PP	—	—	—
PA/PVdC copolymer/PA/polypropylene (Microtort[1])	<2	<2	<2
PET/PA-EVOH-PA co-extrusion/polypropylene	3	9	57
EVOH co-extrusion/polypropylene	3	18	139
PA-EVOH-PA co-extrusion/polypropylene	4	8	37
Amorphous PA/PA/polypropylene	6	17	32
Amorphous PA type 11/polypropylene	2	8	107
PET/SiO_x-PET/CPP	<1	<1	5.5
Camclear HB/polypropylene	<1	<1	<1

[1] Trademark of Lawson Mardon Flexible.

Table 6.5 Pouch flexible packaging materials for in-package microwave processing. OPP and PE inner-sealant layers are shown for possible pasteurisation use only. (Source: Foodpack Information Centre, Leamington Spa, UK.)

OPA/SiO/PE	Met. PET/PE
OPA/PVdC/PE	PET/PVdC/PE
OPA/PVOH/PA/PE	PET/SiO_x/PP
OPP/SiO_x/OPP	OPA/CPP
PET/OPA/CPP	PET/amorphous PA/CPP

passive container, the microwave field needs to be focused to allow maximum opportunity for penetration through the container walls and base. Sealing 'like' materials can result in a permanent bond in which case openability is not easy and cutting with a knife may be necessary. More likely, the packaging will include a peelable element in its sealing layer. This is also the food-contact layer so the needs of food-contact packaging legislation are implicit in any peel/seal packaging developments for foodstuffs. Many of these peel/seal systems use proprietary knowledge that the producers are reticent to disclose. Despite this, general indications of the materials can quickly be analysed using standard techniques such as the infrared examination of polymer spectra.

Today's technology can provide multilayered, flexible packaging structures with barrier properties close to those of aluminium-foil-containing equivalents. For practical purposes, measuring the overall package barrier properties could be a realistic choice. Many potential examples are available where these 'non-foil' structures are helping to maintain shelf-stable shelf-lives of 12 months or longer. While it is true to say that, as yet, hardly any of these products are being processed using microwave technology, there seems no reason to doubt their potential suitability. Current processing involving these materials uses conventional thermal techniques to achieve the desired shelf-life, although it has to be said that product quality levels are generally poor—hence the search for improved thermal stabilisation technologies capable of producing sterile or extended-life products.

Some of the commonly used, sealed-in-place, lidding structures use the same combinations as those for thermoformed containers. However, the individual layer gauges are thinner since no drawdown (and hence sheet-thinning) occurs. The emergence first of metallised aluminium-coated films is now being followed by other inorganic metal oxide coatings, as shown in Tables 6.4 and 6.5. It needs to be emphasised, however, that these coatings are easily damaged and must be carefully handled. To protect their barrier function, they are currently used when sandwiched in flexible laminates between two other supporting films.

6.4.2 *Microwave-active and microwave-reflective materials*

One of the biggest problems associated with the microwave processing of multi-component meals is the need to obtain a uniform temperature distribution throughout each component. Some form of differential heating may be needed if the performance of each component is to reach the desired level. The packaging system chosen must be capable of achieving optimum product quality, sensory quality (texture/taste) and freshness among all of the ingredients of a multi-component meal. This can be helped by developments in the areas of microwave-active and microwave-reflective materials.

Microwave-active field-management devices are well known as 'susceptors' and sometimes as 'receptors'. These can vary the field intensity at given positions on a package. They are made from inorganic metal oxide or nitride coatings (3–400 Å). The most widely used example is metallised polyester film, which is laminated to a cartonboard backing for physical support and to prevent the carrier film from curling as its temperature rises. Others include silicon oxide, silicon nitride and aluminium oxide.

Susceptors are designed to absorb microwave energy and so heat directly in a microwave field. This heat is passed to the foodstuff using a good thermal contact or a radiative heat path. Transferring heat to the product helps prevent the susceptor from overheating. The best results are where the transfer distance is between 0 and 3 mm. According to George *et al.* (1991) raised sections of a package should not be used with receptors. This is because microwaves obey the inverse square law and rapidly decrease in strength over very short distances. A calculated balance is needed between the size of the food, the susceptor and the energy absorbance characteristics of the susceptor.

In the same way that light is reflected by a mirror, most metals reflect microwaves rather than being heated by them. This property is used both for industrial microwave processing (pasteurisation, sterilisation and tempering) and for the heating of food in microwave ovens. Reflection helps the microwaves travel directly to the food. This is why the interior walls of microwave ovens and industrial processing vessels are made of metal. With microwave ovens, a metal grid or wire mesh in the door has a similar reflective effect to that of the metal sheet lining the oven cavity. The door also prevents microwave seepage from the oven. A proportion of the reflected waves are back-reflected on to the microwave generator, the so-called magnetron. In ovens made before 1980, the tops of the magnetron tubes were sheathed by a glass dome. Intense heating in these ovens led to premature failure of the magnetron tubes. This was solved by coating the tubes with ceramics. Arcing is another avoidable problem. Switching on an appliance where a plate with gold edging or a metal container is against or close to an oven wall can produce a spark. Theoretically, the magnetron could be damaged but up to now no damage has been known to occur. The arc is passed between two points where reflected microwave rays in the form of energy have concentrated. The effect has been likened to the charges that show as sparks when electrostatically-charged clothing is touched (Alasuisse, 1988).

Packaging materials that reflect microwaves (principally, uncoated metal trays) can thus be used with care in a microwave oven. They should be placed as near the centre as possible to avoid any possible contact with the walls.

Microwave reflective systems such as 'ameliorators', 'shielding devices' and so on, are used to lower the field intensity at given locations on a package. They

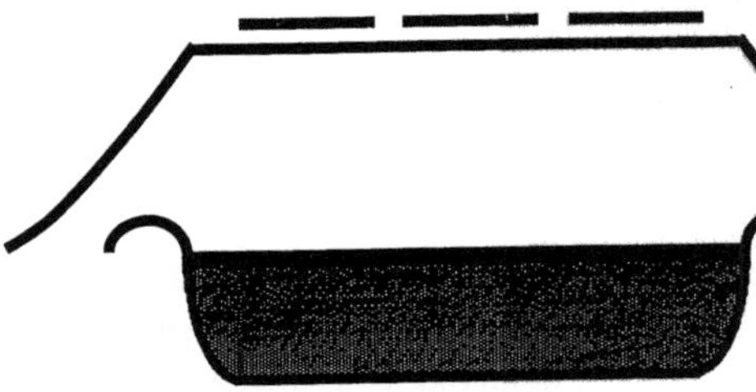

Figure 6.2 Microwave-active field management system. An aluminium foil grid-shaped antenna in the package snap-on overcap is used to focus microwave energy into the product. (Source McCormick, Alcan, Toronto.)

involve the use of metal containers or metallic foils and are at the forefront of prepackaged product cooking or reheating by consumers but have, so far, played a lesser, or even non-existent, role during industrial microwave processing. Nevertheless, this is one of the areas of continued interest and opportunity if uniform heating profiles are to be achieved. Their use could be as an inherent design component of the sealed package or form part of the snap-fit lidded closure (Figure 6.2) usually present over the sealed container. Either role could result in improved thermal processing performance when dissimilar components of multi-component foods are present.

6.4.3 Package geometry for even heating

Heating foods evenly is the goal in all industrial microwave food-processing applications. For many reasons, even when using microwave-passive (transparent) materials, this may not occur. The compromise is to select shapes that allow the coldest parts of prepackaged, processed foods to reach the minimum required temperature in the shortest possible time while not overheating the remainder.

Package shape is also influenced by the nature of the product and both the *rate* and *method* of filling, and also sealing and container handling. These can involve a compromise. For 'conveyorised' handling of continuous systems, square or rectangular tray shapes are simpler than elliptical trays or circular bowls/pots. Double- or triple-compartmented packages are also being used in microwave processing but require a higher degree of care than single-compartmented containers.

Depth-to-diameter ratios are especially important. The more these exceed 1:1 the more potentially unsuitable they become. Some manufacturers persist in using these so as to avoid a head change on their filling and sealing equipment. The use of 125 ml and 250 ml pots with the same filling and sealing aperture is one example of poor choice. (To overcome any reheating shortcomings, consumers are required to frequently stir the product.)

A pronounced cavity on the base of containers is found to give microwave processing advantages, according to Coluccio (1991). Shute (1990) notes that 'packages should have some flexibility. When first introduced into the pressure vessel, cold product makes the packages deform inwards. After heating, the packs

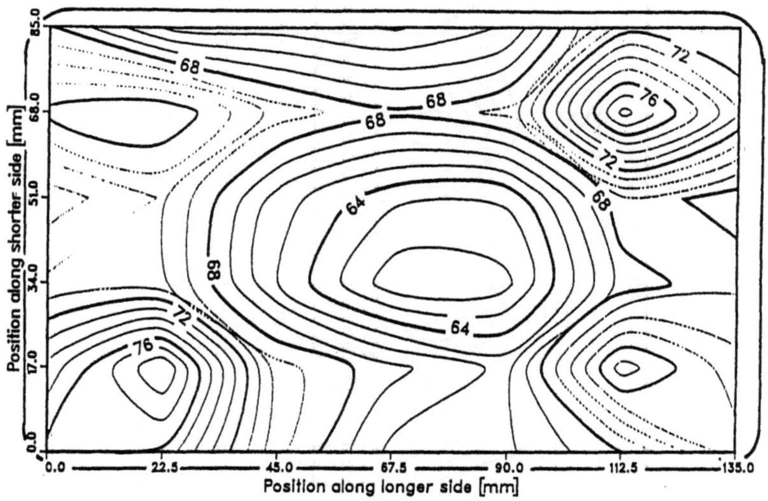

Figure 6.3 Thermal contour plots showing the temperature distribution for microwave-heated foods in rectangular trays. (Source Carnaud Metalbox Technology Centre, Wantage, UK.)

return to near their original ambient shape'. These features take advantage of the inherent flexibility of thermoplastics materials and are incorporated in the package designs currently in use in Europe for microwave food processing.

Researchers at the CarnaudMetalbox Technology Centre (Wantage, UK) have studied three of the main container shapes filled with test food products (Child and Maskell, 1990). Figures 6.3, 6.4 and 6.5 show the food temperature 'contour maps' for each container type respectively described below.

1. Rectangular 300 g tray: a 20°C difference was found in the food temperature. Hot spots were close to the four sharp, angular, corners. According to CarnaudMetalbox, microwave heating comes largely through the electrical standing waves 'bouncing' around in the angle of the corners. This produces extra heating. The corners also expose a large surface area leading to runaway heating effects taking place. Thus trays with such sharp corners are not an ideal shape for microwave applications.
2. Quartic 300 g tray: CarnaudMetalbox developed this tray taking into account the deficiencies found with the rectangular tray. The food map for this tray shows a much closer overall temperature difference of 10°C. The highest temperatures are at the centre, in contrast to the sides or edges as found with the rectangular tray.
3. 8 oz (approx. 225 ml) Aseptic pot: this container exceeds the 1:1 drawdown ratio. It is deep drawn and, admittedly, designed for use with aseptically processed products. These are filled into the pot under aseptic conditions. CarnaudMetalbox carried out microwave reheating tests, which partly explain the explosive, volcano-type eruptive heating effects encountered in domestic microwave ovens.

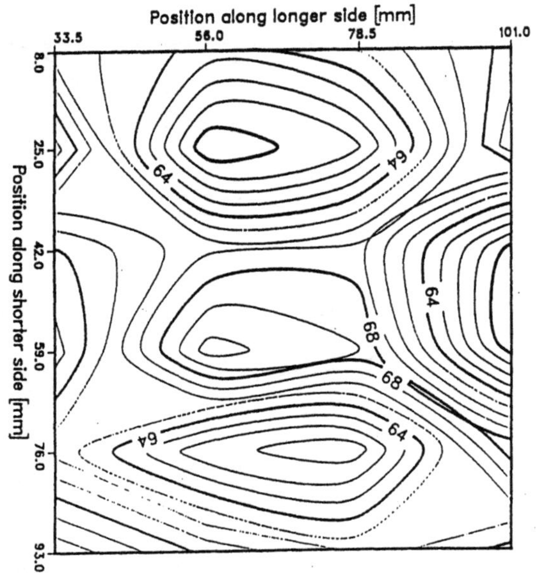

Figure 6.4 Thermal contour plots showing the temperature distribution for microwave-heated foods in quartic (rounded-square shaped) trays. (Source Carnaud Metalbox Technology Centre, Wantage, UK.)

The food map varies tremendously. A temperature difference of over 40°C was found with individual temperatures ranging up to 80°C and higher. It is thought that microwave energy, penetrating deep into the food, creates a superheated region 2–3 cm below the surface of the food. This core heating effect is thought, according to CarnaudMetalbox, to be due to the product acting as a lens to focus the

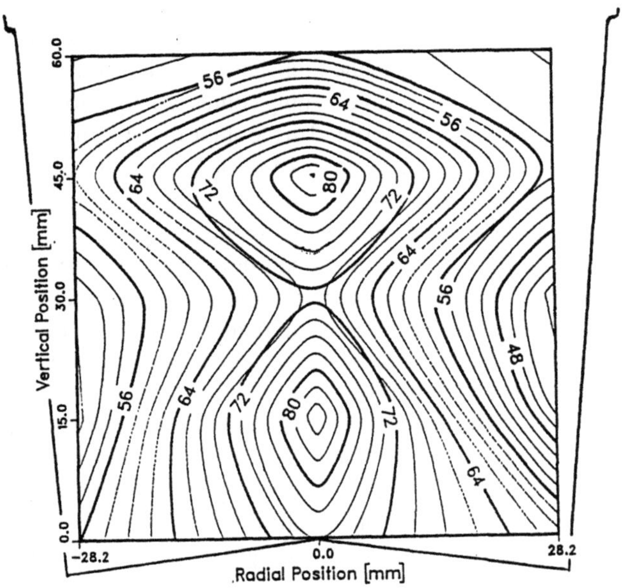

Figure 6.5 Thermal contour plots showing the temperature distribution for microwave-heated foods in 8oz aseptic pots. (Source Carnaud Metalbox Technology Centre, Wantage, UK.)

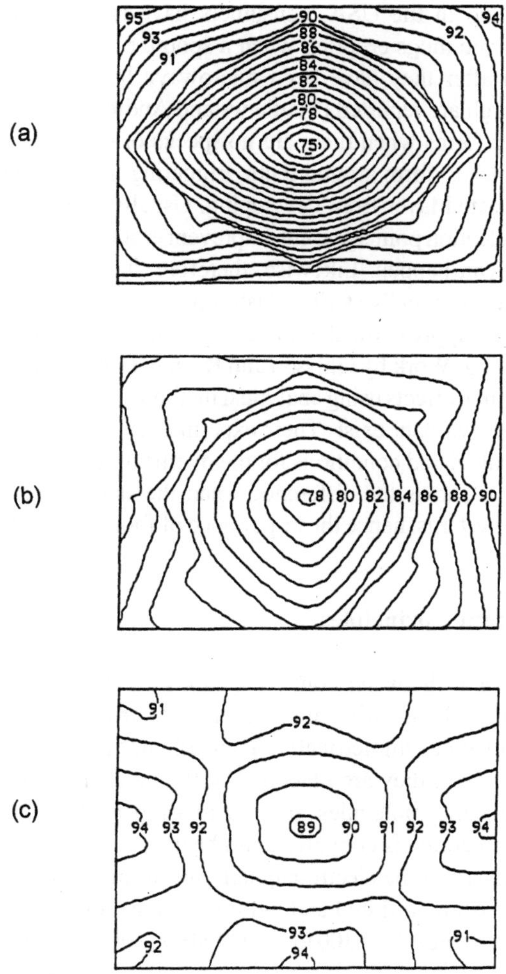

Figure 6.6 Thermal contour plots showing the temperature distribution for the same pre-packaged ready-to-eat meal when heated in three different microwave systems: (a) domestic oven at 2450 MHz, 0.6 kW for 158s; (b) pilot-scale multimode tunnel at 2450 MHz, 6kW for 158s; (c) pilot-scale single mode tunnel at 896 MHz, 7 kW for 144 s.
The peak temperatures were measured 11mm above the base of ready-to-eat meals by the top thermocouple on each probe after microwave heating. (Source Burfoot and James, Institute of Food Research, Bristol, UK.)

microwaves on a specific location. Core heating is especially likely to occur with high-viscosity products. The induced localised heating can lead to the predicted explosive discharge of product. As such, this is definitely not a container that is suitable for prepackaged microwave process applications.

Figure 6.6 shows three similar food maps of the same ready-to-eat meal after heating in three separate microwave systems. Only the product heated in a single mode oven has an approximately uniform temperature. This can be helped by using waveguides to focus the applied microwave energy. These are not used with domestic or industrial multimode microwave systems.

Both in Europe and in the USA, consumers have reported hot beverages in 'squeaky-clean' ceramic mugs erupting when a spoon, biscuit or even a sweetener pill has been added after they have been microwave heated. Researchers at the University of Pittsburgh are reported (Matthews and Andrews, 1991) to believe that the superheated layer forms when there are no specks or cracks in the mug. These normally allow bubbles to form around them, triggering the 'boiling' phenomenon that stops the superheated layer from developing. Usually, there are sufficient dust particles, or other nucleation points present, to prevent eruption occurring. When they are absent, eruption can happen. Likewise, therefore, it is easy to visualise them occurring in dust-free, aseptically sterile, plastics pots.

Edge heating effects apply particularly to foods being microwave processed, with resultant uneven heating. Work by Ohlsson and Risman (1991) suggests that it is not possible to calculate such effects using analytical methods, since virtually no recorded qualitative or quantitative data exist. Their experimental results are now available together with some calculated conditions for the overheating of cylindrically shaped food loads. Figure 6.6 shows the difference when the same package is heated by differing microwave systems, hence the importance of food maps to food processors.

6.5 Microwave heating principles

It is the dipolar nature of the water content of foodstuffs that plays the major role in microwave heating. The positive and negative electrical charges on the water molecules are positioned asymmetrically. They attempt to align themselves with any electric field in which they are placed. The field alternates many millions of times per second. As the molecules attempt to keep up but fail to do so, the microwave energy generated is converted into heat.

Microwaves form part of the electromagnetic spectrum. This also includes radio and television broadcasting, infrared and ultraviolet radiation and visible light. The microwave frequencies range from 300 to 30 000 MHz (300–30 000 million cycles/s)

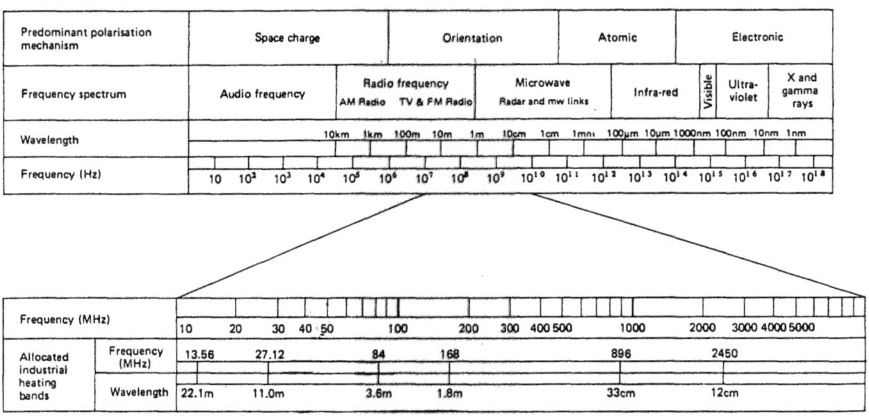

Figure 6.7 Industrial heating frequencies in the electromagnetic spectrum. (Source British National Committee for Electroheat (BNCE).)

compared with the 50–60 cycles/s for domestic electricity supplies. It can be seen from Figure 6.7 that microwaves are positioned close to FM radio and television broadcasting bands. Each location corresponds to both a frequency and wavelength. The simplified version of an electromagnetic wave shown in Figure 6.8 also shows the wavelength. This is the distance between each crest or trough and similar neighbouring positions of a complete wave-cycle form (Andrews, 1991). Likewise, the frequency is simply the number of times/s that the wave passes through one cycle, in other words the time to travel between two adjacent, similar points along the wave. The two are related as follows.

$$\text{Wavelength } \lambda \text{ (m)} = \frac{\text{Velocity (m/s)}}{\text{Frequency(Hz)}} \tag{6.1}$$

6.5.1 The significance of the wave

It can be seen from Figure 6.8 that the electromagnetic wave comprises two elements. The electrical, standing element (E) and magnetic, horizontal, element (H) are at right-angles to one another. It is the former that is of greatest use in microwave-heating applications. The magnetic component assumes prominence where shielding and certain ferro-magnetic browning effects are involved. The amplitude of the two components of the wave (E and H) represents the strength of the respective electrical and magnetic fields.

6.5.2 Microwave interactions with materials

Schiffman (1988) explains that 'Microwaves are a form of energy, not a form of heat'. They are only manifested as heat when they interact with a material following an energy transfer mechanism. The microwave power absorption of a material can be readily expressed:

$$P_V = k.f.E^2.\ \varepsilon'' \tag{6.2}$$

where P_V = power developed in a volume of material, k = constant (to determine the units of power and volume), f = operating microwave frequency, E^2 = electric

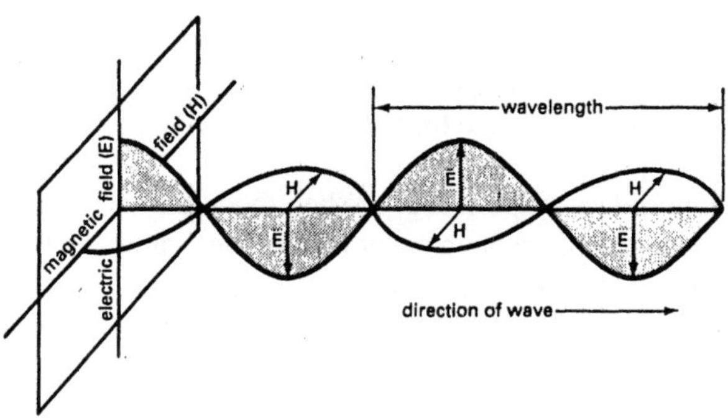

Figure 6.8 Simplified electromagnetic wave representation. (Source Gordon Andrews, Gamma Consultants.)

field strength in volts per unit distance and ε'' = dielectric loss factor for the material. The dielectric loss factor indicates the degree to which an externally applied electric field will be converted into heat.

The rise in temperature follows the laws of thermodynamics:

$$Q = M.\ s.\ t \tag{6.3}$$

where Q = heat required, s = specific heat of the material and t = temperature rise.

The rate of increase in temperature can be expressed as follows:

$$\frac{k'.f.E^2.\varepsilon''}{\rho.\ c} \tag{6.4}$$

where k' = constant expressing heating in desired temperature and time units, f = operating microwave frequency, E^2 = electric field strength, ε'' = dielectric loss factor, ρ = density of the material and c = specific heat of the material.

The frequency and electric field are properties of the microwave system while the density, dielectric loss factor and specific heat relate to the material being heated. Thus there is an obvious interaction between the material being heated and the microwave equipment providing the heating.

6.5.3 *Microwave heating factors*

When materials are heated by microwaves, the process is affected by several factors. These relate to both the microwave system and the material to be heated itself and each plays a part. For prepackaged foods inside microwave-passive packaging, the main factors are:

1. Starting product temperature.
2. Size.
3. Shape.
4. Density or consistency.
5. Specific heat.
6. Thermal conductivity.
7. Dielectric properties.

Where multi-component products are to be processed, these factors are unlikely to be constant. Geometry and consistencies will vary, as will the start conditions, where some ingredients will range from fresh or frozen to part-cooked.

6.5.3.1 Starting product temperature. When foods are heated by microwaves, the temperature rise is dependent on many factors concerning the food. The initial temperature of the food is important in determining the heating rate and heating time. In general, with any heating process, the higher the starting temperature the faster food will cook. This ruling applies to microwave heating. Other important factors that affect the temperatures obtained include latent heat (ice in frozen foods changes to water and additional thermal energy is needed), the dielectric properties and the thermal properties of the food.

6.5.3.2 Size. When, for example, particulate foods are to be heated, items of the same size should heat evenly; smaller particulates will require less heat than larger ones.

6.5.3.3 Shape. The geometry of the foods is similarly important. Overheating can occur with irregularly shaped foods. Rounded, regular, shapes will tend to heat more evenly than particulates with sharp corners or with varying, thick and thin, sections. However, as George *et al.* (1991) point out, a sphere or similarly curved surface may have its hotter regions towards the centre. They suggest that dramatic overheating is not noticeable for radii exceeding 0–50 mm.

6.5.3.4 Density or consistency. Most foods are not of uniform consistency. This affects the way they heat, particularly the evenness. Denser foods will tend to take longer than those that have a more open, porous, composition.

6.5.3.5 Specific heat. This is one of the fundamental properties governing the heating of foods. Specific heat capacity is defined as 'the quantity of heat required to raise the temperature of unit mass of substance by 1 degree centigrade'. It is a measure of the ability of a material to hold heat compared with water. The units of specific heat capacity are J/g/°C.

The specific heat capacity of water is 1.0. For most fats, the figure is about 0.5. Only half as much heat is needed for a given amount of a fat compared with the same weight of water. Specific heat varies with temperature—particularly at subzero conditions. An abrupt change occurs around the freezing point of foods. This reflects latent heat effects since additional heat is needed for a food during the change of physical state between ice and water. Thus additional amounts of energy (as heat) are needed if frozen foods are being microwave-processed to overcome these latent-heat effects. Where the products have reached ambient temperature, however, specific heat far outplays the dielectric loss factor as foods are heated (Figure 6.9).

6.5.3.6 Thermal conductivity. If optimum heating conditions are to be achieved, the thermal conductivity of each component being processed must be known. Overheated surface layers but cool interiors can result from applying too much heat to the surface of foods with low thermal conductivities. The optimum rate of heat supply must match the thermal conductivity of the food if microwave heating and conductive heating effects are to balance (George *et al.*, 1991). A comparison of the specific heats and thermal conductivities of common foods is given in Table 6.6.

6.5.3.7 Dielectric properties. These govern how well a material absorbs microwaves. They are very important to food manufacturers who are formulating foods

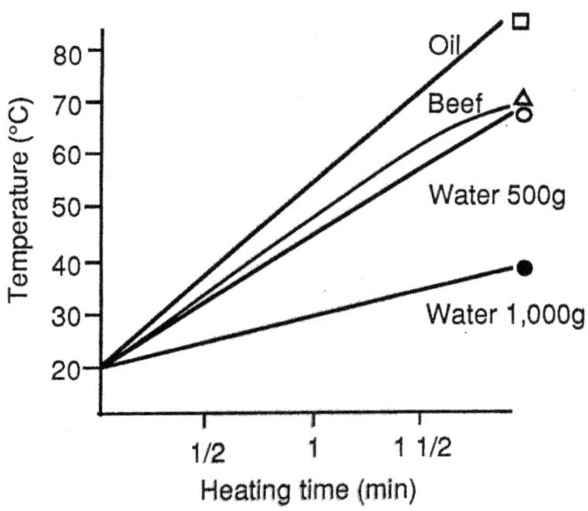

Figure 6.9 Relative heating rates of water, beef and oil, illustrating the dramatic effect of specific heat overshadowing dielectric loss. (Source *Food Technology* (June, 1986), from Ohlsson (1983).)

for successful microwave-process applications. There are three relevant dielectric properties:

1. Dielectric constant, ε′.
2. Dielectric loss factor, ε″.
3. Dielectric loss tangent, tan δ.

Schiffman (1975) notes that, in general, the higher the numerical value of these terms, the better a material will heat. 'The better it heats, the lossier (i.e. more heat absorbent) it is.'

The dielectric constant is a measure of the ability of a material to store microwave energy. The dielectric loss factor shows the ability of the material to dissipate this energy as heat. When characterising the microwave heating of foods, a third useful property is the dielectric loss tangent. This is a measure of the ability of a material to be penetrated by microwave energy, and to dissipate this energy in the form of microwave heating. It is defined as the ratio of dielectric loss factor to dielectric constant. Following the pan-European 'Cost 90 bis' programme, a comprehensive reference list of the electrical and dielectric properties of foods was researched and published (Kent, 1987). Data on a wide range of foods are covered.

Dielectric properties vary according to the composition of foods: carbohydrate, fat, protein, salt and water content all play their part. Figure 6.10 shows the relative microwave-absorption properties of some foods indicating the effects of temperature, dielectric loss properties and salt content. Temperature also directly affects the dielectric properties. Experimental studies (Bengtsson and Risman, 1971) show how the dielectric constant varies with temperature—there is a large difference between thawed and frozen foods, for example. The work at Campden Food and Drink Research Association on the layout of multicomponent foods is helping to resolve some of the uneven heating conditions that result from the runaway heating effects

Table 6.6 The thermal properties of foods. (From Schiffman, 1976, and King, 1989)

Food	Specific heat capacity (kJ/kg.°C)	Thermal conductivity (W/m.°C)
Meats		
Lean	3.06–3.43	0.36–0.61
Fat	2.59–2.89	0.20–0.26
Bacon	2.01	
Soups	3.09–4.10	—
Vegetables		
Fresh	3.77–4.10	0.30–0.42
Dried	1.84	0.21
Fruit		
Fresh	3.73–4.10	0.40–0.55
Dried	2.14–2.47	0.3–0.4
Milk		
Whole	3.85	0.54
Skimmed	3.98	0.57
Cream	3.14	0.57
Miscellaneous		
Bread	2.85	0.35–0.45
Egg white	3.85	0.45–0.50
Egg yolk	2.81	0.33–0.97
Butter	2.09	
Vegetable oil	1.46–1.88	0.23

often found as frozen foods are reheated. Partly-thawed food preferentially absorbs more of the available microwave energy. A rapid temperature rise occurs in these part-thawed regions. Meanwhile, the frozen parts of the food attract much less energy. The result is that some parts reach the desired temperature while others are still frozen.

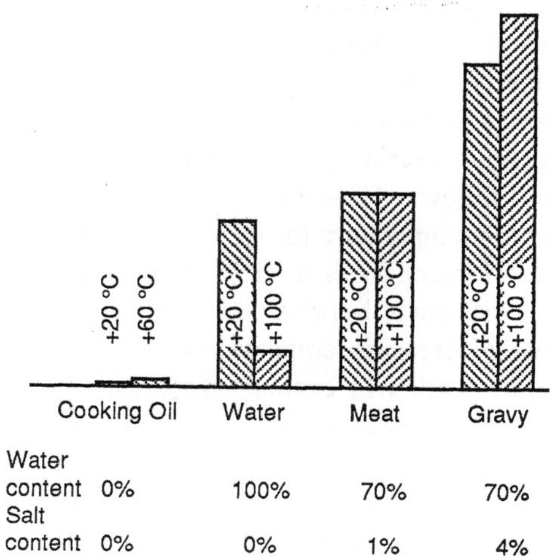

Figure 6.10 Relative microwave absorption properties of some common foods showing the effects of temperature, dielectric loss properties and salt content. (Source *Food Technology* (June, 1986), from Ohlsson (1983).)

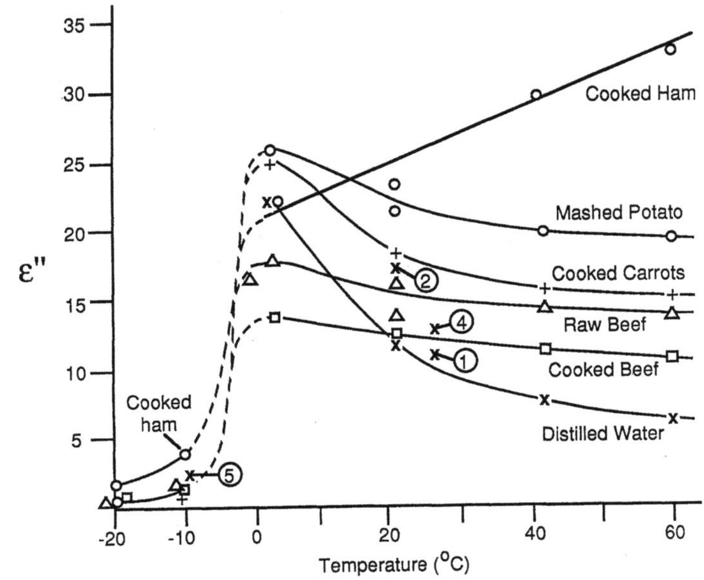

Figure 6.11 Temperature dependence of dielectric loss factor (ε") for several common foods at 2.8 GHz. (Source Bengtsson and Risman (1971).

These effects are represented in the data shown in Figure 6.11.

Freezing has a major effect on the way materials heat because of the very different dielectric properties of ice and water. Water is highly absorptive and heats easily, whereas ice is highly transparent to microwaves and heats poorly. Thus microwave thawing can produce the partial runaway heating described. Hence, microwave tempering is preferred. Here product temperatures are raised to just below the thawing point after which conventional processing or reheating (microwave or otherwise) follows. These effects are depicted in the way the dielectric loss factor varies with temperature for various food products. It increases rapidly as the temperatures of frozen foods rise towards 0°C but then begins to fall. Figure 6.11 shows that cooked ham is the only food examined where the dielectric constant continues to rise with increasing temperatures.

Campden's future work will evaluate the effects of changing the dielectric properties on the microwave heating behaviour of foods. It is known, for instance, that the surface of foods can be altered to absorb microwave energy better. However, while the surface heats better, the microwave penetration depth decreases and the interior of the food is inadequately heated.

The dielectric properties of foods and other materials at 2450 MHz and at 20°C are shown in Table 6.7.

Table 6.7 Dielectric properties of foods and other materials at 2450 MHz at 20°C. (From Schiffman, 1975)

Material	Dielectric constant	Dielectric loss tangent
Distilled water	78	0.16
Raw beef	49	0.33
Mashed potato	65	0.34
Cooked ham	45	0.56
Peas	63	0.25
Ceramics	8–11	0.0001–0.001
Plastics	2–4.5	0.001–0.02
Glasses	approx. 4	approx. 0.001–0.005
Papers	3–4	0.05–0.1
Woods	1.2–5	0.01–0.1
Ice	3.2–3.3	0.0007–0.001

6.6 Extending product shelf-life

It has always been good business practice for food manufacturers to continously search for new methods and/or technologies that are capable of extending the shelf-life of foods while ensuring that products giving high yields, high operating efficiency and high quality are made at the lowest or optimum cost. The traditional ways of preserving foods include salting, smoking, drying, canning/bottling plus chilling or freezing. Each of these has a drawback: most of the accepted traditional techniques result in a lowered perception of freshness of the particular food product involved. Refrigerated or frozen foods may be perceived to be closer to fresh product but involve additional costs through the need to maintain the cold chain throughout the 'unopened' life of the prepackaged product. Hence, extending the shelf-life of high-quality foods (with good retained sensory levels while avoiding as much of the cold chain as possible) remains a high priority for food manufacturers.

6.6.1 HTST processing

Seen from the viewpoint of food processing research and development, the most interesting aspect of the microwave revolution is the possibility of using microwave technology for the preservation of foods. From the late 1970s onwards, the then state-of-the-art thermal processing and preservation technologies involved scraped-surface or tubular heat-exchange systems.

Today, thermal processing development activities are being driven by the desire to improve the quality of food products. This especially applies to the integrity of particulates and the reduction in overcooking by using high-temperature/short-time (HTST) processing. Cost constraints increasingly make it necessary to reduce or optimise the costs of packaging materials and containers. Hence any likely possibility of utilising new types of packaging must be given serious consideration. Fortunately, the packaging industry has produced some fairly dynamic advances throughout the 1980s to match these needs and to overcome the technical hurdles encountered along the way. Commercial pressures have seen these advances continue to emerge whether through totally new single materials or combinations of new and existing ones.

Aseptic processing, using scraped-surface or tubular systems, has shown itself to be successful as an in-process stabilisation technique capable of handling both high- and low-viscosity fluids (e.g. beverages, stews, simple dessert puddings, pour-over sauces and so on) including many with small particulates. An important limitation, however, with both aseptic processing and the assembly of those products that contain large particulates is the ability to maintain the particulate integrity. This applies equally to the processing operation and subsequent passage of the product from the processing equipment to the filler and thence into the packaging container or system.

6.6.1.1 Benefits of HTST over traditional processing. Ever since Nicholas Appert became the inventor and father of thermally processed foods (following the publication of his '*Appertisation*' process in 1810) the art of preserving foods by heat (rather than drying) has been increasingly refined. The product descriptions may vary (with canned, bottled, tinned and potted food and so on) but Appert's process to provide shelf-stable foods has largely remained unchanged until now, with the arrival of aseptic processing techniques.

For decades, retort technology has been used for commercial food processing. It has one big drawback. The resultant processed foods do not have a fresh taste, appearance or quality. Retort technology can safely process food, even those containing large particulates, and produce sterile products. However, high temperatures and long processing times are necessary. The combined effect of these factors is a loss of colour, flavour, freshness, nutrient value and texture. Despite these drawbacks, this technology is still one of the most widely used today, even for the many shelf-stable or extended-life products designed for microwave reheating by consumers.

Pasteurising in a retort, at lower conditions, is an alternative thermal-stabilisation technique used to produce foods with higher quality than those that have been

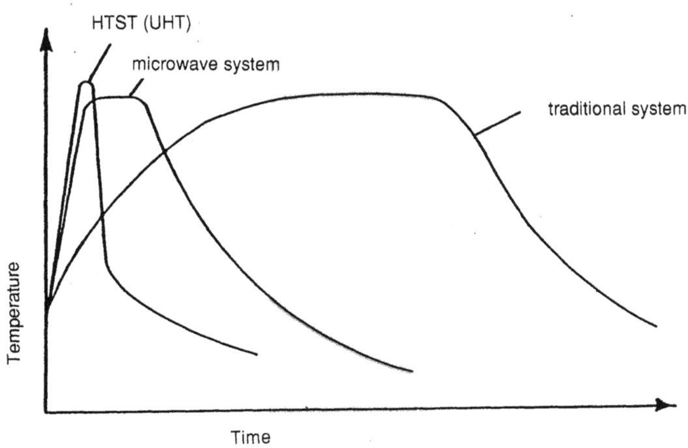

Figure 6.12 Typical time-temperature heating and cooling profiles for foods under competing thermal stabilisation systems (traditional retort, HTST and microwave processing). (Source OMAC (1991).)

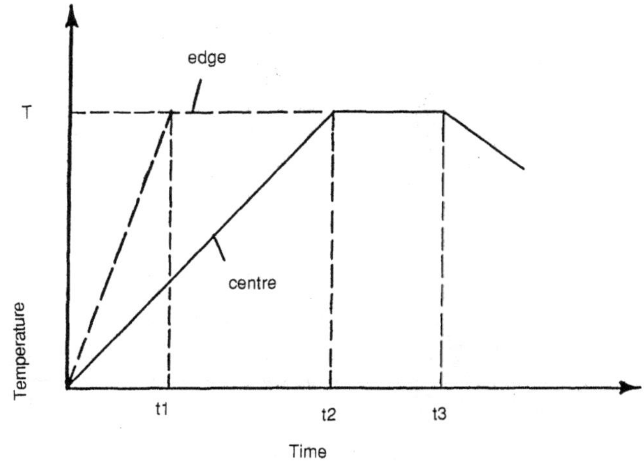

Figure 6.13 Time-temperature profile for a traditional retort system showing wide differences between the outer edge and core product temperatures. I = treatment temperature; t_1 = edge heating time (5 min); t_2 = equilibration time (20–25 min); t_3 = holding time (3–10 min). (Source OMAC (1991).)

retort-sterilised. The drawback is that pasteurising only results in products with a limited shelf-life extension. Hence, the opportunities presented by aseptic processing and packaging to provide superior quality products are seemingly vast. Microwave processing, through its potential ability to handle large particulate-containing foods, is at the forefront of the thermal stabilisation technologies capable of producing sterile products. It shows much promise, although some of the first commercially produced products have not been of the quality expected.

Figure 6.12 shows the time–temperature heating and cooling profiles during traditional retort, HTST (UHT) and microwave processing of foods. Figures 6.13 and 6.14 respectively show the claimed differences in time–temperature profile

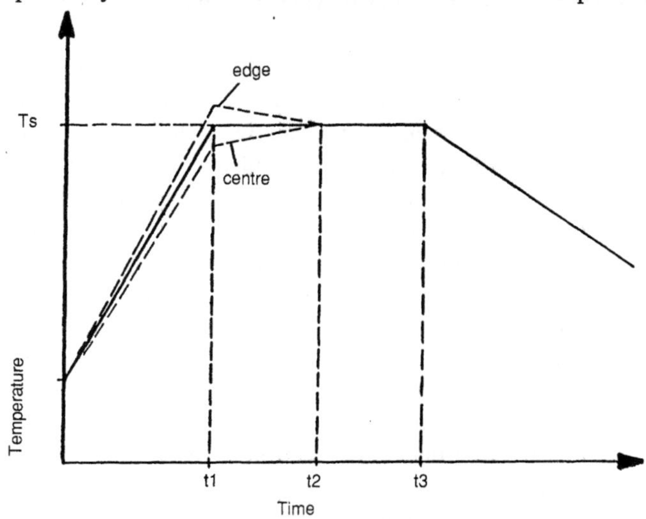

Figure 6.14 Time–temperature profile for a microwave system showing a relatively uniform temperature throughout the product. T_s = set point temperature; t_1 = heating time (5–7 min); t_2–t_3 = equilibration time (3–5 min); t_3–t_2 (3–10 min).

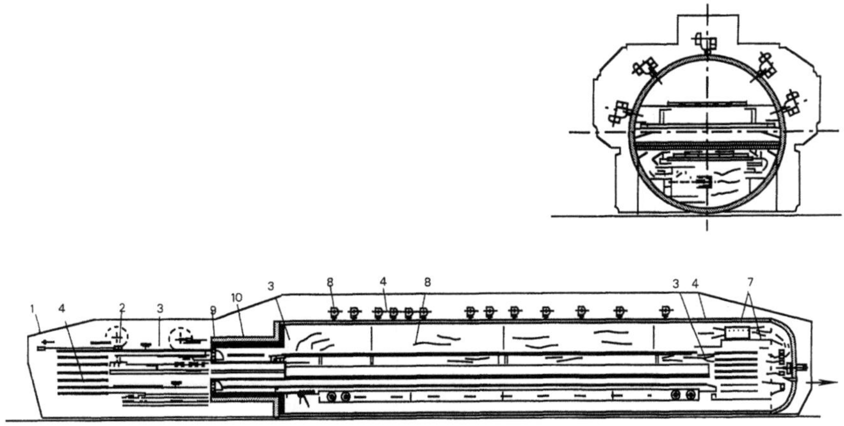

Figure 6.15 Pressurised, continuous 2450 MHz microwave sterilisation system with entry and exit from the same end of the pressure vessel. Microwave heating takes place as products pass along the upper conveyor. Trays of products then descend and return through a lower cooling section. Cooling can be with refrigerated air or iced water. (Source OMAC, ReggioEmilia, Italy.)

between the edge and centre of a food during retorting and microwave processing. According to one microwave equipment manufacturer, Omac of Reggia Emilia in Italy, microwave processing is capable of producing a uniform temperature distribution throughout the product.

6.6.2 *Microbial and sensory decay*

Two mechanisms have a significant effect on the overall appearance, colour, flavour and taste. These are microbial contamination and organoleptic deterioration. Industrial microwave-processing techniques have been shown to inactivate enzymes, retard or eliminate microbial growth and prevent organoleptic breakdown of the food components. However, microwave processing, while no longer in its infancy, is still not a widely developed process despite some limited use in Europe, Japan and the USA. For nearly 20 years, the challenge has been to prove that microwave processing can be used effectively to pasteurise and, latterly, to sterilise various types of foods including, in particular, multicomponent particulate-containing products.

If an industrial microwave oven is to be effectively and efficiently used as a thermal-processing device, the appropriate parameters and process conditions for pasteurisation and sterilisation must first be determined. Pasteurisation inactivates vegetative organisms; sterilisation inactivates spore-forming microorganisms. Decareau (1991) observes that there is considerable concern about the microbiological effectiveness of the microwave-sterilisation process. Very few data are reported in the technical press. One exception comes from Mudgett and Schwartenberg (1982) in the USA, but much of their work is predictive rather than practical. In Europe, Ronchetti (1989) opts only to provide microbiological information on pasteurisation, despite his company being one of those producing continuous microwave-sterilisation systems (Figure 6.15).

Microwave pasteurisation is now being used in Europe, Japan and the USA to extend the life of bread, pasta and ready meals. Processing occurs in a non-pressurised

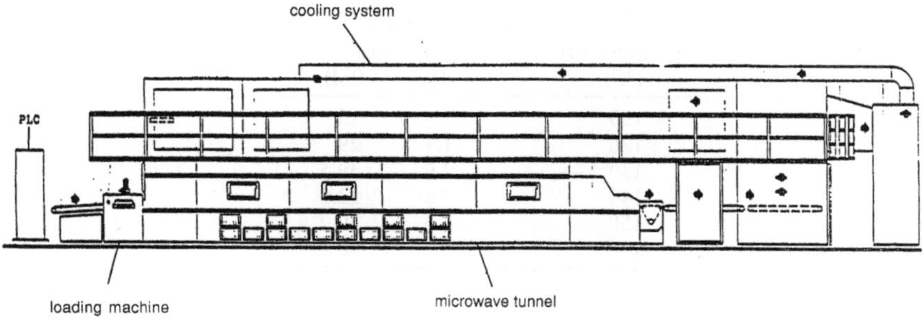

Figure 6.16 Microwave pasteurisation system for bakery, meat and pasta products. (Source OMAC, ReggioEmilia, Italy.)

microwave chamber or tunnel; the temperatures involved are below the boiling point of water. Decareau (1991) reports that factors attributable to the effect of high-frequency heating on microorganisms contribute to the lethality of microwave processing, in addition to the destruction caused by the generation of heat in the food products in the standard time–temperature relationships (Figures 6.16 and 6.17).

6.6.3 Pasteurisation versus sterilisation

According to Shute (1990), pasteurisation and sterilisation differ in terms of temperature, which is, in turn, dependent on the type of product to be processed and shelf-life sought. To pasteurise milk requires around 30 min at 63–65°C. Similarly, meat products may, typically, require 5 min at 75°C while packaged bread products could need 5 min at 90°C. In Europe, microwave pasteurisation at temperatures between 74°C and 85°C is also being used to extend the shelf-life of pasta products by companies in several countries.

Figure 6.17 Microwave pasteurisation system for ready-to-eat meals in temperature-resistant plastics trays. The line has four treatment chambers each equipped with four 1.2 kW generators (total 57.6 kW). (Source Hermann Berstorff, Maschinenbau, GMBH, Hannover, Germany.)

Table 6.8 Absolute pressures to raise products to boiling point. (From Shute, 1990)

Temperature (°C)	Pressure (Bar	/ PSI)
100	1	15
120	1.7	29
130	2.6	39.5
140	3.5	52.5
150	4.6	70
160	6	90

All of these processing times and temperatures can vary according to the type of product, the amount being processed, packaged product geometry and the 'holding' time at a given, requisite processing temperature. As the temperature rises, the processing time can be shortened.

Microwave sterilisation involves higher temperatures. All are above the boiling point of the water present in the foods. It is not possible with commercial thermal processing to raise the water temperature above 100°C without evaporating it off as steam, while the prepackaged product remains under ambient processing conditions. However, raising the pressure allows higher processing temperatures to be reached and sterilisation to be achieved. In addition, very little water is lost (Table 6.8).

Very few pressurised microwave sterilisation systems have yet been commercially developed except for unpackaged foods. With prepackaged foods, the packaging materials chosen must withstand the high temperatures and internal pressures generated by the expanding headspace gases (principally steam). A balance is needed between retaining the integrity of the package and ensuring the requisite thermal treatment has been achieved.

Using a pressurised heating chamber prevents the rupture of the package seal and explosion of the package as the headspace gases expand.

The package should also not distort during processing. If, for example, a package retains its integrity during processing but emerges physically (and visibly) distorted, then not only will any high-quality appearance be lost, but the consistency of the product will also probably be lowered. There must also be a strong likelihood of a distorted package not passing through an automated packaging system without encountering further damage. Two package shapes used for microwave-sterilised foods are shown in Figure 6.18.

Another key factor in any microwave pasteurisation or sterilisation process is the need to establish a uniform temperature distribution throughout the food product without the development of isolated 'hot spots' or 'cold spots'. These are one of the most common hurdles to be overcome during microwave processing. Clearly, the shallower the container and, hence, prepackaged product profile, the better will be the chance to avoid a big thermal gradient in the product both during and at the end of processing. In order to determine the magnitude of this thermal gradient, the appropriate instrumentation needs to be provided and used.

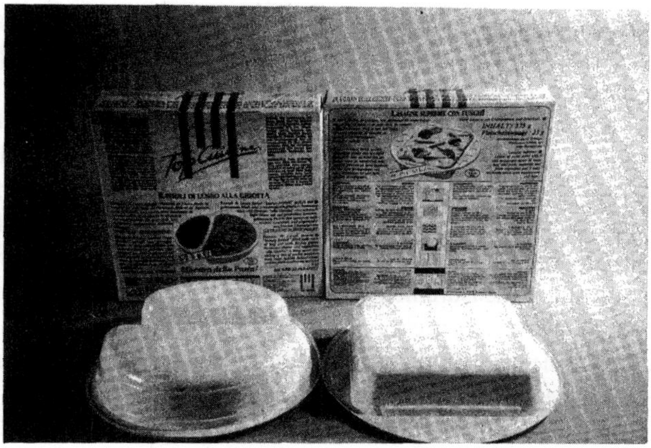

Figure 6.18 Microwave-sterilised pasta ready-to-eat meals. The multilayered PP/EVOH/PP trays used illustrate a design compromise. To enable the food manufacturer to have only mimimal filling line changes, the same plastic trayforming moulds are used. These have interchangable inserts to produce either shape but with a common overall size.

6.6.4 *Temperature measurement*

Owing to their interference with the magnetic field, conventional temperature-sensing devices, such as thermocouples, cannot be used. Their accuracy is impaired and there is a real risk of arcing occurring at the interface between the sensor and the target. This makes it unrealistic to consider using thermocouples in a microwave chamber. The problem is exacerbated where packaged products are being conveyed through a tunnel inside the chamber rather than remaining in a static position.

Since large-scale commercial production involves conveyorised movement through the microwave chamber, at least two alternative temperature-measurement systems are being used commercially. Fluoroptic thermometry (Luxtron Corp., 1988) can accurately determine the internal temperature of products undergoing microwave heating. The optical fibre probes used are non-metallic and are thus not affected when placed within a microwave field. Another technique is to use infrared pyrometry (Labfacility Ltd, 1980; van de Voort *et al.*, 1987). This gives a profile of the surface temperature distribution and can be combined with a fibre-optic system to provide a more detailed assessment of the overall thermal gradient. Outside of the pressure vessel, thermocouples can be used. It is not uncommon for over 30 thermocouples to be used in a so-called 'hedgehog' array to measure both the surface temperature of prepackaged processed foods and the temperature at any desired point or depth within the container.

One problem yet to be solved by fluoroptic systems is that, like thermocouples, they also involve the use of 'leads'—this time of optical fibres. This means that, while they can safely be used inside the microwave chamber, they still have to trail back to the temperature 'measuring' unit. This is acceptable for periodic quality checks and research testing in a static mode but not for continous conveyorised, in-line production. Infrared pyrometry is thus the main method used inside the microwave processing unit, with fluorometry available for more detailed information support.

Infrared thermal-imaging techniques are also fast becoming established as the means of providing a visual representation of the thermal profile and, perhaps more importantly, to show the positions of 'hot spots' and 'cold spots'.

Fluoroptic systems (Luxtron Corp., 1988) use probes with a tiny speck of luminescent phosphor embedded in the tip. This responds to pulsed blue light sent along an optical fibre by emitting a bright-red fluorescence. The rate of decay of its intensity is determined by sensors and calibrated against known data to find the temperature. It is almost certain that each major food and equipment manufacturer will be using fibre-optic temperature-measurement systems. However, they are expensive, cannot be used for continuous-flow processing inside the pressure vessel and are very fragile. Careful handling is needed. In the static mode, it is possible to measure inside the pressure vessel, since the leads can be up to 2 m in length.

In addition, ultrasonic devices are in the process of being developed (Redwood, 1960; Meeker and Meitzler, 1964) as a means of measuring temperatures during microwave applications. This is because sonic waveguides are similar to electromagnetic waveguides. As such, they cause waves to propagate in several different nodes. Experimental evidence (Nicholson and McDicken, 1985) suggests that these can be available in the form of either continuous or pulsed waves. The ultrasonic devices under development are claimed to show promise as a means of determining the thermal profile without the use of connecting wires. The technique has been developed at least to the demonstration model stage (Richardson, 1989).

Other information needed must allow the dielectric properties and, separately, specific heat, of the food products to be determined. This is especially important where multicomponent products are to be processed. Instrumentation is available to enable data to be determined for each individual food component. The information can be used both to increase knowledge and in the creation of a fundamental database to predict the behaviour of foods being microwave-processed.

6.6.5 Microwave sterilisation

When microwaves are used to sterilise a food product, several unit operations are necessary to achieve a thermally stabilised product. These largely correspond to those used in conventional thermal-processing operations. The following are essential for complete thermal stabilisation:

First-stage process: heat the product to sterilisation temperature.
Second-stage process: hold the product at the sterilisation temperature.
Third-stage process: cool the product to below the boiling point.

The pressure can be up to 4 atmospheres (4 bar) in order to achieve the necessary temperatures of 121.1°C or above.

One of the biggest problems still largely unsolved is achieving a holding step within a pressurised microwave processing unit (see Fig 6.19). It is the holding step that maintains the product at the sterilisation temperature for a predetermined time period to ensure microbial destruction is complete. However, it is difficult, when

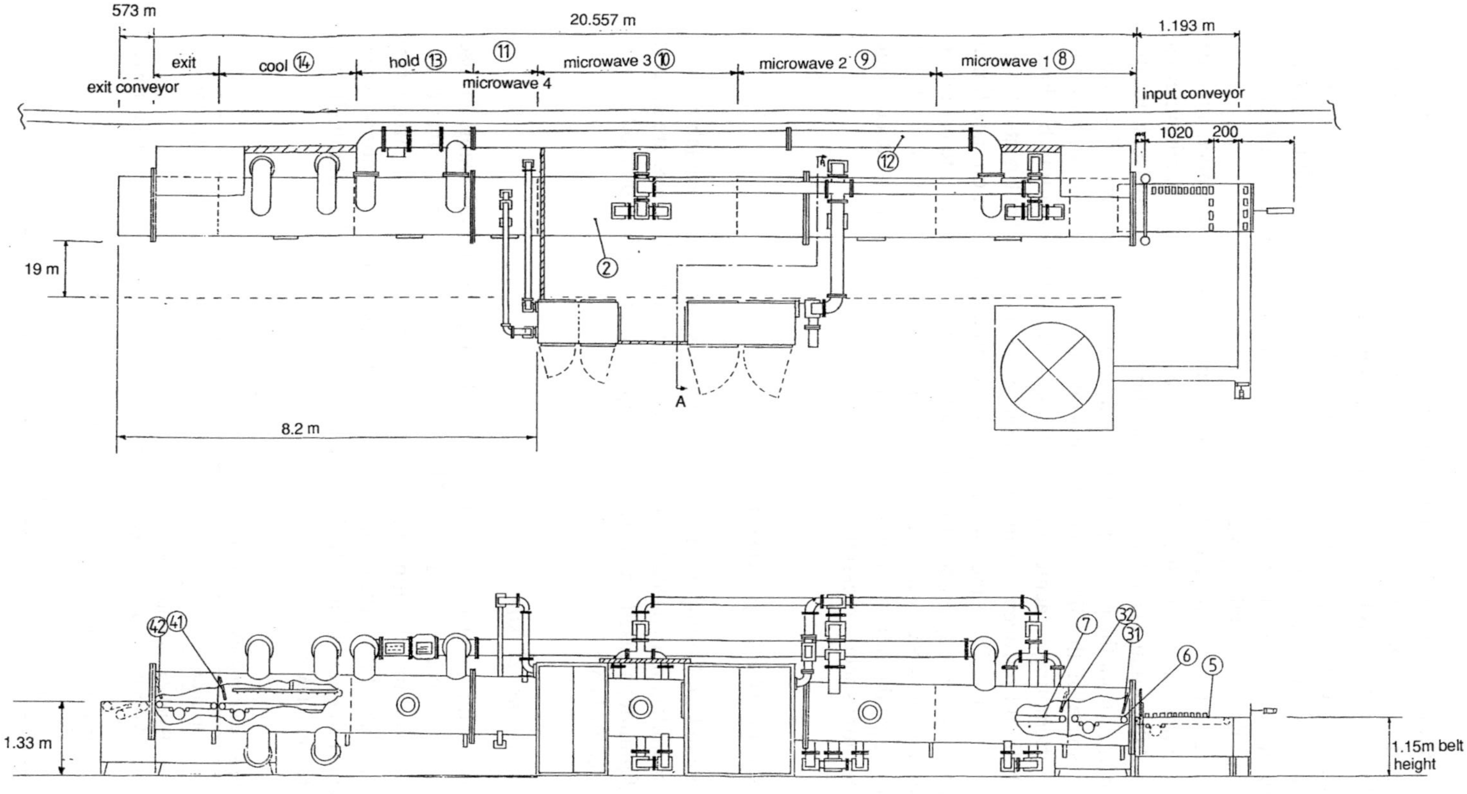

Figure 6.19 Side elevation and plan view of a pressurisable, in-line, continuous 915 MHz microwave system. This unit also has a low-powered 2450 MHz applicator to help avoid any surface cooling effects on the surface of the product.

designing a hold step, to accurately determine and control the degree of microwave energy needed to maintain the sterilisation temperature throughout the product. One thought would be for other types of heat-transfer mechanisms to be developed for use as the hold step during microwave processing. Some surface cooling effects have also been found. To guard against these, one solution (Shute, 1990) is to use a secondary, lower-powered, microwave source. Another (Coluccio, 1991) uses hot air to assist the stabilisation process.

In the design of a microwave sterilisation unit, a cooling chamber must be incorporated immediately after the holding section. This is in order that the internal temperature of the product is reduced below the boiling point. Otherwise, when the package emerges from the pressurised chamber, its internal pressure will cause it to explode. Some of the first commercially available microwave sterilisation systems are attempting to achieve this by using rapid surface-cooling by direct contact with liquid nitrogen, refrigerated cold air or conventional heat-transfer media such as water. Once the product exits the processing section, additional conventional cooling is again possible. This lowering of the product temperature also prevents any possible overcooking and/or loss of sensory quality. It also helps to ensure a safe production operation as the packages of processed product emerge under normal atmospheric pressure.

6.7 Types of microwave processing systems

Both pressurised and non-pressurised microwave systems are available from equipment manufacturers based in Europe, Japan and the USA. There are several design configurations based on operational frequency and desired method of use: these are listed in Table 6.9

6.7.1 The frequency debate

There is some controversy over the selection of the frequencies allowable for use for industrial microwave processing. The equipment radiates energy that could interfere with telecommunication equipment and control. To avoid this, frequency bands were allocated for industrial, scientific and medical (ISM) purposes by the International Telegraph Union (ITU) and agreed by the World Administrative Radio Conference (WARC) in 1979.

Table 6.9 Types of microwave processing systems

Type	Operational mode	Frequency	Wavelength	Area of use
Non-pressurised	Batch (static) chamber	896 MHz	33.48 cm	UK
Non-pressurised	Continuous (conveyorised) tunnel	915 MHz	32.79 cm	USA and Japan
Pressurised	Batch (static) chamber	2450 MHz	12.25 cm	Worldwide
Pressurised	Continuous (conveyorised) tunnel			

Figure 6.20 60 kW generator for 896 or 915 MHz. (Source Microwave Systems, Ampthill, UK.)

British-made equipment is known to be operating safely in continental Europe on both 896 and 915 MHz. This is done by ensuring that the radiation leakage levels are well screened and suppressed to such low levels that external telecommunications equipment is unaffected. However, continental European manufacturers' equipment is largely based on 2450 MHz.

In the UK, 896 MHz will discontinue as a WARC-approved frequency for use; however, in practice, equipment will continue to be made and sold. As in continental Europe, the leakage levels are such as to be shown not to interfere with telecommunications equipment. Cellular radio, potentially one to be most affected, deliberately avoids approaching the 896 MHz band. Microwave processing equipment made in the UK is designed to comply with European interference standards such as CISPR II. This *status quo* looks to have avoided the need for a successful outcome to the current negotiations to find an alternative heating band in the 896 MHz area.

The magnetron electric valves used to generate microwave power differ. Those used in the 896–915 MHz bands are between 25 kW and 60 kW. Magnetrons for 2450 MHz are much lower-powered—usually between 0.5 kW and 5 kW (max.

Figure 6.21 Overall view of an in-line pressure vessel from the exit end. Microwave-processed product emerges on the exit conveyor visible in the centre of the picture. In the background, the 60 kW 915 MHz generator can be seen alongside the pressure vessel (see Figure 6.19 for closer details). (Source Microwave Systems Ltd, Ampthill, UK.)

about 6 kW). One company, APV Baker of Crawley, UK, is understood to be developing a 15 kW 2450 MHz generator. APV already produce low-powered generators for 2450 MHz and high-powered generators for 896 and 915 MHz. A typical free-standing 60 kW generator for 896 or 915 MHz is shown in Figure 6.20. This can also be seen in Figure 6.21 which shows its use alongside a pressurisable 915 MHz microwave system in Europe.

Food manufacturers face the question whether it is preferable to use one or more high-powered magnetrons or a multitude of low-powered magnetrons. For their part, manufacturers of industrial microwave processing equipment (896, 915 and 2450 MHz) are incorporating different types of waveguide and antennae feeding systems to help control, distribute and direct microwave energy towards the product. For some applications, equipment operating at any of the three frequencies can be used for microwave processing applications with a barely perceptible difference in performance. Those products having a small size and reduced thickness lend themselves to processing at 2450 MHz. It may, however, prove impractical to choose 2450 MHz as the operating frequency. This is because difficulties associated with cooling the magnetrons limit the low maximum power available. Thus any controversy surrounding frequency selection should focus on the sensitivity of the product to microwave energy and the operating effectiveness of the equipment.

6.7.2 Penetration depths

Outside of the UK and continental Europe, the majority of industrial microwave-processing equipment operates at 915 MHz and with 25–60 kW magnetrons. At low temperatures, there is more than 2.5 times greater penetration depth at 915

MHz than at 2450 MHz. Thus the frequency of 915 MHz allows the processing of products with increased thicknesses to a higher degree of uniform temperature distribution. As the temperature rises, these differences in penetration depths are reported to narrow (Mudgett, 1986) due to the increase in dielectric loss factor with ionic conductivity at 915 MHz. The penetration depths become closer with increases in product temperature; at the boiling point of water they are approximately equal.

Buffler and Stanford (1991) acknowledge that it is extremely difficult to determine accurately the exact penetration depth value, 'the depth of penetration of a microwave signal into a food product, commonly defined as the depth where the microwave signal has dropped to 36.8% of its value when first entering the product'. Thus a knowledge of the penetration depth is invaluable when preparing thermal contours for food maps.

6.7.3 Operating efficiencies

The efficiency of energy transfer between the microwave processing equipment and the product can also be affected by the product's basic electrical properties. Efficiencies are taken into account when preparing operating costs.

6.7.4 Relative operating costs for industrial microwave processing

When compared with conventional techniques, microwave processing shows several advantages, including efficiencies and the amount of factory space needed. It is necessary, however, to know the costs of operating microwave processing equipment. These vary depending on whether the operating frequency of the equipment is in the 896–915 MHz or 2450 MHz bands. Edgar (1986), McKinnon (1991) and Sanio (1986) all give indicative microwave processing-cost information.

6.7.5 Batch versus continuous equipment

The decision on the type of industrial microwave processing system to use is influenced by several factors. These include the required throughput, processing time and finished-product quality, plus both the effectiveness and efficiency of the process. Increasingly, manufacturers are tending to opt for continuous- rather than batch-processing equipment. This choice is largely governed by a need for significantly lower operating costs and process times. Continuous systems are well established for non-pressurised systems that pasteurise foods. Some of the very latest involve up to 60 different computer programs where multicomponent foods are involved.

6.7.6 Continuous microwave pressure vessels

Continuous pressurised systems are capable of sterilising foods. However, very few have yet been built. Those that have are being monitored and developed further

Figure 6.22 Polypropylene containers for microwave-processed foods showing the peelable membrane and snap-on overcaps. (Source Foodpack Information Centre, Leamington Spa, UK.)

by the microwave equipment manufacturers—often working in partnership with one or more food producers.

One of the first to be used commercially is shown in Figures 6.19 and 6.21. Products are packaged in sealed, microwave-passive, containers (Figure 6.22). These enter the pressure vessel via a rotary valve-controlled airlock. Inside, the vessel is pressurised to balance the expansionist forces inside the package. The pressure stops the containers breaking open during heating. There are three distinct sections inside the vessel: (i) microwave application; (ii) holding; and (iii) cooling sections. Within the microwave section, heating takes place for either pasteurisation or sterilisation as required. In the first instance, this unit is being used commercially in the pasteurisation mode but it can (and has been) pressurised to produce sterilised ready-to-eat meals and snack products. The 60 kW 915 MHz generator used can be seen alongside the pressure vessel in Figure 6.21 and is shown in greater detail in Figure 6.20.

It can be seen from Table 6.8 that this so-called ‘overriding counter-pressure’ prevents the water content of foods boiling when the temperature rises above 100°C to achieve sterilisation temperatures of 121.1°C and above. Eisner (1988) describes in great detail the significance of overriding counter-pressure from the aspects of:

1. Filling factors.
2. Filling procedures.
3. Packaging containers.
4. Sterilisation factors.

After heating to the required temperature in the microwave section, often referred to as the come-up time (CUT), product stays in the holding section where recirculating hot air helps allow pasteurisation or sterilisation to take effect, as conductive heating helps achieve an even temperature profile throughout the product. Either

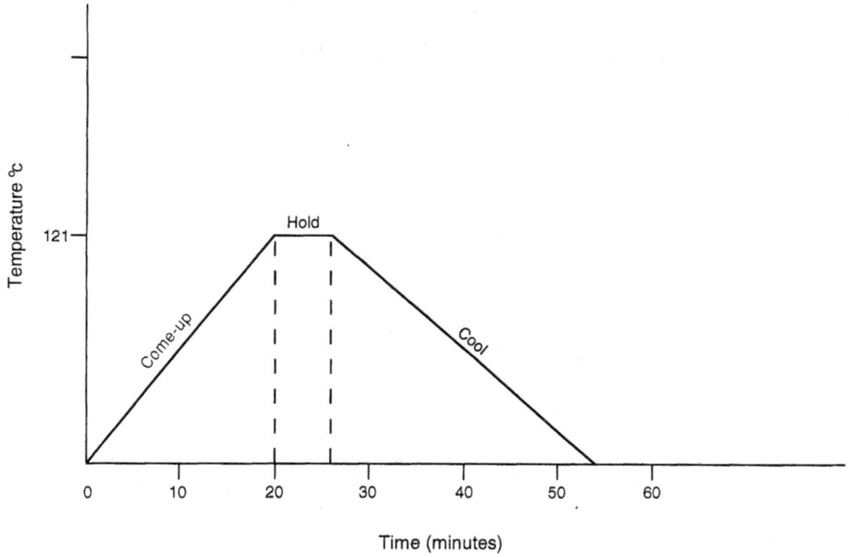

Figure 6.23 Time–temperature profile for microwave sterilisation at 915 MHz showing the come-up time (CUT), holding step and subsequent cooling. (From Microwave Systems Ltd, Ampthill, UK.)

indirect steam heating or electrically heated elements are used to heat the air.

While still under pressure, standard refrigeration techniques are used to cool the product before leaving the pressure vessel via a second rotary valve-controlled airlock. The integrity of the package seal is thus retained and product can be packed immediately for shipment or storage.

Figure 6.23 shows a typical heating, holding and cooling pattern using such industrial microwave processing equipment.

6.7.7 Thermal stabilisation

Industrial microwave processing equipment is usually the only unit in a system for the thermal stabilisation of food products. This is, however, not always the case. It can also be used as the key element in a multicomponent technological system (MCTS). These take advantage of the ability of microwaves to rapidly penetrate into materials, as already seen. Microwave technology can thus be considered for use in combination with other stabilisation systems to achieve the highest possible product quality, freshness and safety.

During the mid-1980s, significant food-processing developments occurred in the use of high-temperature/short-time (HTST) thermal-processing techniques. These have included both pasteurisation and sterilisation using microwave processing systems. HTST processing here uses a combination of microwave and thermal energy to sterilise foods containing large particulates. The technology involved has used both thermoform/fill/seal pouches and shallow, premade, plastics packaging trays (Stensom, 1983). Using HTST processing, sealed packaged product is placed in a circulating water bath, the temperature of which is controlled and equilibrated to the surface temperature of the package during the heating cycle. This procedure helps to produce a uniform temperature distribution throughout the product. There are significant product characteristics associated with this type of processing system:

1. A high degree of retained natural colour.
2. Optimum quality.
3. An extension of shelf-life.

Microwave and packaging technologies are capable of providing and protecting the superior product quality sought by today's consumers. They are already being shown to be highly compatible with in-package stabilisation and extended-life, food-preservation techniques. The concept of creating innovative microwave preservation systems for prepackaged foods still presents one of the greatest challenges to the food producer, packaging materials manufacturer and converter alike, as well as to the microwave systems manufacturer.

A potentially bright future exists for microwave processing and the packaging materials involved but key advances are still needed to overcome the unresolved challenges.

Acknowledgements

Grateful thanks are expressed to the Ministry of Agriculture, Fisheries and Foods, and the American Institute of Chemical Engineers, for information provided in this chapter.

References

Alusuisse Ltd (1988) *Aluminium Foil Containers in Microwave Ovens*. Alusuisse Ltd., Neuhausen am Rheinfall, Switzerland.
Andrews, G. (1991) Safety and performance standards. In *Microwave Ovens*, Peter Perigrinus, London.
Bengtsson, N. E. and Risman P. O., (1971) Dielectric properties of foods at 3 GHz as determined by a cavity perturbation technique. II. Measurements on food materials. *Journal of Microwave Power*, **6**(2):120.
Buffler, C. and Stanford, M. (1991) *The Effects of Dielectric Properties on the Microwave Heating of Foods*. 26th Annual Microwave Symposium, The International Institute of Microwave Power, Buffalo, New York.
Child, H. and Maskell, A. (1990) *Microwave Packaging Update*. Microwave Association Annual Conference, Telford, UK.
Coles, R. E. (1991) Lasers used to make easy-opening lines on packs. *Food, Cosmetics and Drugs Packaging*. Elsevier Advanced Technology, Oxford, England.
Coluccio, C. (1991) *Microwave Pasteurisation*. Ambient Shelf-Stable Products Conference, London, May 1991, Processing and Packaging Machinery Association and the Institute of Packaging, UK.
Decareau, R. V. (1991) *Microwaves in the Food Processing Industry*. Academic Press, New York.

Decareau, R. V. (1991) *Microwave Pasteurisation and Sterilization.* International Microwave Power Institute, 26th Annual Microwave Symposium, Buffalo, N.Y.
Decareau, R. V. and Peterson, R. A. (1986) *Microwave Processing* and *Engineering*, VCH, Deerfield Beach, Florida.
Edgar, R. (1986) The economics of microwave processing in the food industry. *Food Technology*, Inst. Food Technologists.
Eisner, M. (1988) *Introduction into the Technique and Technology of Rotary Sterilisation.* Private Author's Edition, Milwaukee, Wisconsin.
Freedman, G. (1985) *The Historical Development of Microwaves.* Unpublished address to General Foods Corp., White Plains, New York.
George, M., Burnett, S. and Richardson, P. (1991) The Influence of Thermal, Electrical and Physical Properties on the Quality of Foods Heated by Microwaves: Final Report. *Technical Memorandum No. 624*, Campden Food and Drink Research Association, Chipping Campden, UK.
Kent, M. (1987) *Electrical and Physical Properties of Food Materials.* Science and Technology Publishers, Hornchurch, UK.
King, W. R. (1989) Listeria *and* Salmonella *in Perspective.* Proceedings of the Microwave Technical Association Conference, London.
Labfacility (1980) *Temperature Sensing with Thermocouples and Resistance Thermometers—A Practical Handbook.* Labfacility Ltd, UK.
Luxtron Corporation (1988) Accurate temperatures in microwave ovens are difficult to achieve. *Food Engineering* **60** (5):175.
Mathews, R. and Andrews, G. (1991) A dirty mug takes bang out of a cuppa. *Daily Telegraph*, London, November 1991.
McKinnon, R. G. D. (1991) *Pasteurisation and Sterilisation using Microwave Energy.* Shelf-Life '91 Conference, London.
Meeker, T. R. and Meitzler A. H. (1964) Guided Wave Propagation in Elongated Cylinders and Plates. In *Physical Acoustics—Principles and Methods.* Mason W. P. (ed.), Academic Press, London.
Mudgett, R. E. (1986) *Microwave Properties and Heating Characteristics of Foods.* Food Technology, The Institute of Food Technologists, June.
Mudgett, R. W. and Schwartenberg, H. G. (1982) *Microwave Food Processing: Pasteurisation and Sterilisation—A Review.* The American Institute of Chemical Engineers.
Nicholson, N. C. and McDicken, W. N. (1985) *Modelling of Ultrasonic Wave Propagation in Waveguides.* Dept. of Medical Physics and Medical Engineering. Royal Infirmary, Edinburgh.
Ohlsson, T. and Risman, P. (1991) *Heating Patterns of Food Edges and Cylindrically Shaped Food Loads in Microwave Ovens.* International Microwave Power Institute, 26th Annual Symposium, Buffalo, New York.
Redwood, M. R. (1960) *Mechanical Waveguides.* Pergamon Press, New York.
Richardson, P. (1989) *Temperature Measurement in Microwave Ovens.* Internal Publication, Campden Food and Drink Research Association, Chipping Campden, UK.
Ronchetti, R. (1989) *Food Pasteurisation and Sterilisation with Continuous-Cycle Microwave Equipment.* MW Pack, Milan.
Rosenberg, U. and Boegl, W. (1987) *Microwave Pasteurisation, Sterilisation, Blanching and Pest Control in the Food Industry.* Food Technology, The Institute of Food Technologists.
Sanio, M. R. (1986) *The Economics of Microwave Processing.* Masters Thesis, University of Texas, Austin.
Schiffman, R. F. (1976) *Journal of Microwave Power.*
Schiffman, R. F. (1975) *Transactions of the Microwave Power Institute*, Vol. 5, Decareau, R. V. (ed.)
Shute, R. A. (1990) *Pasteurising and Sterilising by Microwave Energy.* Microwave Association Annual Conference, Telford, UK.
Stensom, L. A. (1983) US Patent Nos: (1) 3,668,358 (2) 3,809,844 (3) 3,809,445 (4) 3,814,889; Multitherm unveiled. *Food Manufacture*, **35**(10):23.
van de Voort, F. R., Laureano, M. and Smith, J. P. (1987) A Practical Thermocouple for Temperature Measurement in Microwave Ovens. *Canadian Institute of Food Science Technology*, **20**:279–284.
Lasers used to make Easy-Opening Lines on Packs. *Food, Cosmetics and Drugs Packaging*, May 1991.

7 The Dole process

F. S. WHITE

7.1 Introduction

The Dole aseptic canning system has been a commercial application since the late 1940s. The equipment is the foremost process for low-acid food packaging of aseptically processed food. Through the past 40 years, over 60 systems have been placed into production throughout the world, almost exclusively packing low-acid food in metal cans. The Dole system is currently owned and operated by Graham Engineering Corporation.

7.2 Technology overview

The Dole system packages aseptically processed products in metal cans. The system operates at speeds up to 500 cpm (cans/min) within a sterile machine operating at atmospheric pressure. The Dole system uses superheated steam as the sterilant. No chemicals are used in the process. Sterile product is introduced into the sterile integrity of the Dole system and filled; a sterilized cover is then applied to the can and seamed on with a sterilized seamer to produce a clean, sterile and safe product.

7.3 The Dole system

Metal cans with seam on ends are used in all Dole systems. Can sizes from 4.5 oz, 202 × 214 to a number 10 can, 603 × 700 are compatible with the system. The cans for smaller sizes are generally aluminum, with steel three-piece and two-piece cans used for larger sizes. The ends/lids can be standard or easy-open types.

The cans are manufactured with heat-resistant coatings, common to the can industry, and the lids are made with a high-temperature Plastisol sealant, which is also standard in the industry. The decoration of the can is either 'bright' (undecorated) or with temperature-resistant lithography.

The Dole 'plant' consists of four basic components:

1. The can sterilizer;

2. The filling section;
3. The lid sterilizer;
4. The sealing machine.

These components are linked together with a integrated network of instrumentation, controls and alarms.

Packaging of aseptic particulates has not quite yet become commercial for retail packaging as there are still process issues with the FDA. The Dole equipment packages some product with small particulates such as rice and pieces of peppers and onions less than 4 mm in diameter.

7.3.1 Can-sterilizing unit

The can-sterilizing unit is a stainless steel, double-insulated tunnel. Superheated steam at a temperature of approximately 500°F (260°C) is introduced into the tunnel sections to provide the heat of sterilization. The steam is superheated by electric heaters supplied with the system. (Older Dole systems used gas to superheat the steam.) The cans are moved through the tunnels on high-speed stainless conveyors. The number of tunnels and the speed of the conveyors is varied to establish the overall capacity of the system.

The cans are conveyed in the sterilized unit for approximately 45 s until the surface of the container reaches 430–435°F (221–223°C), which is the temperature point required for the destruction of high-heat-resistant bacteria. The superheated steam is more lethal than dry hot air, and requires less time to penetrate the bacteria.

7.3.2 The filling section

As the hot sterile containers move into the filling section, jets of cold sterile water are directed against the lower portion of the can exterior, markedly reducing the container temperature. At this point, the aseptic product is introduced into the containers by one of several specially designed filling mechanisms, which, when properly adjusted, automatically maintains a constant fill level at speeds up to 500 cpm.

The Dole system is equipped with a specially designed filler. The basic model is the slit filler. The cans pass underneath a slit opening in a tube-type filler. The slit is about 1/4″ × 6″ (6 × 152 mm approx.) long, although the size is variable to the type of product and speed of filling. The cans pass along the length of the filling ribbon of product. The tight formation of the cans forms an overlapping flange, thus product waste is less than 0.5%.

The slit filler consists of a thick-walled stainless steel pipe that is slit along its lower side (Figure 7.1). This pipe has a second pipe inside, which has multi-parts and provides consistent flow to the filling system. The annulus between the pipes is about 0.5″ (12 mm approx.). The sterile product is fed into the inner pipe and flows out of the slit in the outside pipe.

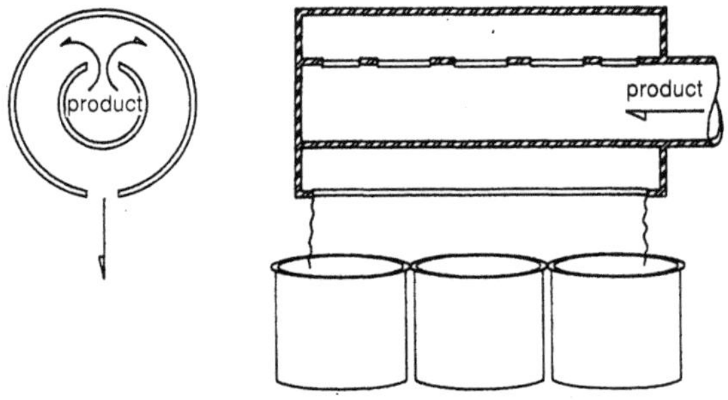

Figure 7.1 Design of slit filler.

The third basic component of the Dole aseptic canning system, the cover sterilizing unit, is arranged as part of the closing machine. Covers are fed in stacks into the unit through one of several mechanical devices, and are completely enveloped in superheated steam at temperatures between 450°F (232°C) and 500°F (260°C) for approximately 75–90 s. At the discharge the covers are completely sterile and are conveyed into the closing machine. Both aluminum and steel lids can be used in the unit. Aluminum lids require about 20% less time. Easy-open lids can also be used in the unit.

The fourth and last component of the system, the closing machine, is usually standard machinery modified to operate in an aseptic canning system. To accommodate the sterile containers filled with cold sterile product and the seaming of the presterilized containers, the closing machine is enclosed and heated with superheated steam. This enclosure permits initial sterilizing of the equipment and maintains a completely sterile atmosphere in the area where the sterile container, cover and product are exposed. The continuous flow of superheated steam creates a positive pressure inside the can seamer, preventing external bacterial-laden air from entering.

Conventional can seamers from CCC, Angelus and Canco are used. The can-closing process takes place under superheated steam and results in a vacuum build-up in the can. To prevent excessive vacuum in the can, which could cause can collapse, either sterile air or nitrogen can be blown into the can immediately prior to closing.

The seamer operates at elevated temperatures of 350–425°F (177–218°C), which necessitates the use of high-temperature-resistant food-grade lubricants for the chucks and seaming rolls. These lubricants are generally commercially available throughout the world.

7.3.3 Instrumentation and controls

In addition to the four basic components described above, there is a complex system of instrumentation, controls and alarms that allows the process to perform properly at high speed. It is imperative from a bacteriological standpoint that accurate temperature readings are maintained at critical points in the process. Thermocouples in these critical areas transmit continuous temperature readings, which are detailed on a strip chart-recorder to become a permanent record of the operation. Visual temperature devices are also included as a support system. Should the temperature in any part of the system fall below a predetermined critical level, both visual and audible alarms are activated. If corrective action is not taken by the operator, the equipment can be automatically shut down, and the containers, covers and product automatically bypassed. The equipment remains inoperable until proper corrective action has been taken.

7.3.4 *System operations*

7.3.4.1 Presterilization of system. The Dole system principle is to presterilize the entire operating equipment prior to start-up. It is essential to sterilize all product- and package-carrying components, including the product filler, the can sterilizer tunnel, the filling chamber, the cover sterilizer and the seamer (Figure 7.2).

1. The first step in the sterilization cycle consists of a hot-water sterilization of the product systems, the Dole filler valves, the can cooling, wash sprays and vacuum reduction, if supplied with the machine.
2. The second step in the sterilization cycle is to heat the Dole canning unit with superheated steam, combined with heating the filling nozzle with the superheated steam. A minimum temperature of 350°F (177°C) must be maintained on all parts for at least 20 min. This procedure sterilizes all

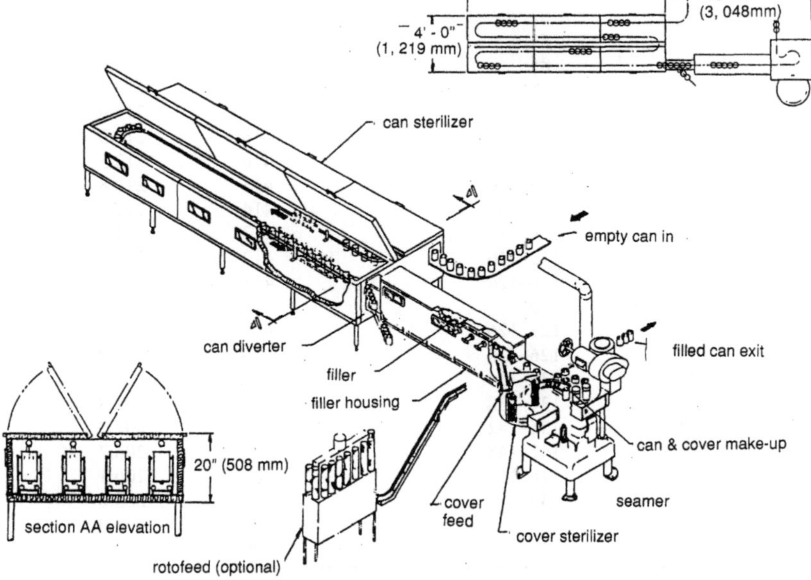

Figure 7.2 Layout of Dole processing line.

Table 7.1 Lethality date for sterilizing empty cans (From Dole Aseptic Canning)

Time in seconds for containers of:		At a temperature of:	
Aluminum	Tinplate	°F	°C
24	30	450	232.2
26	32	447	230.5
27	34	445	229.4
29	36	444	228.9
30	38	443	228.3
32	40	441	227.2
34	42	439	226.1
35	44	438	225.5
37	46	436	224.4
38	48	435	223.9
40	50	434	223.3
42	52	433	222.8
43	54	432	222.2
45	56	431	221.7
46	58	430	221.1
48	60	429	220.5
50	62	428	220.0
51	64	427	219.4
53	66	426	218.9
54	68	425	218.3
56	70	424	217.7
58	72	424	217.7
59	74	423	217.2
61	76	422	216.7
62	78	421	216.1
64	80	420	215.5
66	82	420	215.5
67	84	419	215.0
69	86	418	214.4
70	88	417	213.9
72	90	417	213.9
74	92	416	213.3
75	94	416	213.3
77	96	415	212.8
78–83	98–104	414	212.2
83–89	104–112	413	211.7
89–93	112–116	412	211.1
93–96	116–120	411	210.6
96	120	410	210.0
98–101	122–126	409	209.4
101–104	126–130	408	208.9
104–107	130–134	407	208.3
107–110	134–138	406	207.8
110–114	138–142	405	207.2
114–118	142–148	404	206.7
118–123	148–154	403	206.1
123	154	402	205.6
125	156	401	205.0
126	158	400	204.4

internal parts of the Dole machine, and prepares the machine chambers for the actual operations.

3. The third step consists of adjusting all temperatures of the machine and filling valves to operating temperature conditions. The can sterilizer section is operated at 425–450°F (218–232°C) as is the cover sterilizer. The temperature in the filling and can-closing sections can be operated at 300–350°F (149–177°C) (Table 7.1 and Figure 7.2).

7.3.4.2 Operating conditions. At completion of the three-step sterilization cycle, the equipment is ready for the canning operations. The temperatures at all 11 central points are checked on the control system and the chart-recording system (used to produce a tracking record of the process) is engaged. The empty cans are fed into the can sterilizer, the covers are fed into the magazines of the cover sterilizer and the product filling valves are opened.

7.3.4.3 System resterilization. Resterilization is required whenever the temperature, at any point in the system, falls below the specified safe minimum temperature as set forth by the Dole Process Authority. Resterilization must follow the three-step sterilization procedure.

7.3.5 Superheated steam

The Dole system uses superheated steam to condition and maintain sterility of machinery and sterilize the can and lid during the system operation.

Historically, the Dole system used natural gas heat to superheat the steam. The steam was passed through stainless steel pipes that were located above the open gas flames. Temperature controls were used to moderate the gas flow and the steam temperature. The superheated steam then was discharged through manifolds and pipes into the various sections of the machinery.

In 1986, Dole revised the superheating concept and introduced an electrical superheating alternative, which also allowed the can-sterilization system to be redesigned for a more heat-efficient operation. The electrical superheating system is comprised of 7–11 subdivided units that provide heat to the specific sections of the machinery. For example, the seamer, cover sterilizer and filling sections have individual electrical heaters, thus providing specific heat for each of these components. The can sterilizer section uses from four to seven heaters for the zones within the can sterilizer. The difference in number of zones is related to the overall size and capacity of the system.

The design of the can sterilizer was changed by decreasing the amount of tunnel space required. The tunnel is now smaller in cross-sectional area, thus reducing the overall size. The new design incorporated easy to open panels, view sights, and double insulation, and all were designed into the stainless steel structure. The types of Dole models, their operating speed and can sizes used are shown in Table 7.2.

Table 7.2 Dole models

Model	Can size (fl.oz.)	Operating speed (cpm)
330	18–96	100–70
520	4.5–46	225–124
530	4.5–46	338–145
540	4.5–46	450–155
1210S	4.5–46	60–23
1220L	18–96	40–30
1305	Development line	5–10

8 Microbiological aspects of aseptic processing and packaging

D. A. JOYCE

8.1 Introduction

Microorganisms are an integral part of our environment. They are present in the air we breathe, the water we drink and the food we eat. Soil contains about 0.5 tonnes of living microbial biomass in the top 25 cm of a hectare of pasture land and fruit and vegetables sustain large populations of microorganisms. Microbes are an essential component in the decay of organic matter and the recycling of nutrients. It is not surprising, therefore, that some species are responsible for the spoilage of food.

Microorganisms have evolved and adapted to every conceivable ecological niche and many exist in a benign symbiotic state with a host plant or animal. They exist in vast numbers in, and on, the human body and other animals. In some cases, this colonisation can cause adverse bodily reactions and the microorganisms cause a disease of the host such as an infection or an intoxication (poisoning). Food may be the vehicle for the carriage of a pathogenic organism or may be contaminated with microbial toxins and foodborne infections or food poisoning ensues.

Food deteriorates in four ways: (i) biochemically, such as by oxidation; (ii) by the action of enzymes naturally present in food; (iii) by microbiological deterioration; and (iv) by physical changes, such as the loss or gain of moisture. It has been stated that aseptic technology is used to extend the shelf-life of foods by the destruction of microorganisms and enzymes within the food. Microbial and enzymic activity are two factors that cause raw, untreated food to deteriorate. With these two components eliminated, it is the chemical deterioration of the food that determines the shelf-life of the product (Gill, 1990). In microbiological terms, aseptic processing and packaging of food has the two objectives of, firstly, controlling the numbers of microorganisms in food by a suitable process and, secondly, preventing the recontamination of the food after processing with microbes, which would subsequently grow and render the product unpleasant or harmful.

Aseptic thermal processes can be combined with other methods of extending the shelf-life of products including: (i) preservatives; (ii) oxygen-scavenging chemicals to control oxidation; (iii) vacuum, gas-flushed and modified atmosphere packaging; as well as (iv) choosing packaging materials with regard to their oxygen and moisture permeability. Such considerations are beyond the scope of this chapter. Nor will the chapter consider the detailed operation of different types of aseptic processing equipment but will concentrate upon the operational principles required to control or eliminate unwanted microorganisms and the prevention of recontamination, and outline a quality assurance approach to safe food that is free from spoilage and with the required keeping qualities.

8.2 Foodborne microorganisms

In general terms, the production of a food product of the desired quality from a range of raw materials requires the prevention of microbial contamination and the control of microbial multiplication. The food is processed for safety and recontamination is prevented.

The term 'microorganism' is in common usage and can cover any microscopic organism within the categories of algae, bacteria, fungi, protozoa and viruses. In the context of food hygiene and food poisoning, it might also include other invertebrate parasites such as tapeworms. Representatives from all these groups can cause food-related problems as outlined below.

1. *Algae*: some are responsible for the production of toxins, for example certain aquatic dinoflagellates, which are taken into fish and shellfish. Algae rarely cause problems after processing.
2. *Bacteria*: these are the main category of organism responsible for food spoilage and food poisoning. A number of genera are involved including, but not limited to, *Salmonella*, *Vibrio*, *Clostridia*, *Staphylococcus*, *Bacillus*, *Escherichia* and *Campylobacter*.
3. *Fungi*: these are responsible for spoilage and the production of toxins (mycotoxins) the best known of which are aflatoxins. Yeasts are of significance due to their ability to ferment sugars to produce gas, alcohol and other by-products.
4. *Viruses*: the viruses and sub-viral particles are being increasingly recognised as causing food-related illness. More studies are required.
5. *Protozoa and other parasites*: these are relatively minor problems in the UK but of much more significance in other parts of the world. Parasitic involvement includes problems such as amoebic dysentery.

However, most problems specific to aseptic processing and packaging of food are due to bacterial and yeast contamination, with occasional filamentous fungal problems. Other microorganisms may contaminate raw materials and the organisms, or their toxins, may survive the cooking process.

Other microorganisms are of uncertain significance, in particular, information on the effects of viruses in food is incomplete. There has been interest in recent times in the possibility of prions being transmitted from diseased animals into the food chain. The term prion is derived from 'proteinaceous infectious particle' and is postulated as a replicating protein material containing no nucleic acids. Such an 'organism' has been suggested as the causal agent of scrapie—a disease in sheep—and, more significantly, as the agent of bovine spongiform encephalopathy (BSE). Fears have been expressed that this agent can survive many forms of heat treatment of food and pass into the human population. At present there is little evidence to support this view, despite the concerns expressed by the media and general public.

This chapter will concentrate on bacterial problems associated with aseptic processing and packaging.

8.2.1 Food poisoning

The manufacturer of aseptically processed foods is concerned with several facets of microbiological activity. Firstly, there are the beneficial aspects of microbial growth, typified by the production of yogurt by bacteria or alcoholic beverages by yeasts. Of more general concern are the problems of food spoilage, caused by a wide range of microorganisms given the right conditions. For a food to show evidence of organoleptic changes due to microbiological activity it is probable that populations in excess of a million (10^6) microorganisms per gram of food have colonised the substrate. A yogurt, which after all is only an example of controlled spoilage, will contain at least 100 million (10^8) bacteria per gram. In general, food spoilage occurs when there is microbiological contamination of the product plus favourable conditions for the multiplication of those contaminants.

'Food poisoning' is a term understood by most people to mean the occurrence of diarrhoea and/or vomiting with or without a fever brought on by the consumption of food that is 'off'. It is normally associated with the ingestion of food that is contaminated with pathogenic bacteria and/or their toxins, although it should be remembered that some foods, such as red kidney beans or toadstools, contain natural toxicants, and chemical contamination or adulteration can cause poisoning.

Food can also be the vehicle for the carriage of certain diseases, such as cholera (caused by *Vibrio cholerae* biotype *eltor*), although this is not normally regarded as food poisoning *per se*. In the most widely accepted use of the term 'food poisoning' it is possible to distinguish two forms: (i) infections of the intestine (foodborne infections); or (ii) the action of toxins produced by microorganisms prior to consumption of the food or in the intestine (foodborne intoxications or toxic food poisoning). It is wrong to assume that a food contaminated with food-poisoning organisms or toxins will show signs of spoilage as, frequently, contamination can occur with no sign of organoleptic changes to the food. The food processor must also be fully aware of the possibility that the food may carry low numbers of pathogens, which would be insufficient to cause clinical symptoms but which could multiply at some later stage in the product life-cycle, especially under uncontrolled conditions in the home, and cause food poisoning. The objective has to be to produce food free from pathogenic microorganisms and their toxins.

Typical food poisoning organisms include:

1. *Clostridium perfringens*: *C. perfringens* is frequently found in meat products. This bacterium produces heat-resistant spores found in dust, human

and animal faeces. Food poisoning is caused by the infection of the intestine where an enterotoxin, which damages the intestinal lining, is released.

2. *Clostridium botulinum*: this is an anaerobic, spore-forming bacterium that produces a highly poisonous toxin that affects the nervous system and may result in death due to paralysis of the respiratory centres. The spores are resistant to heat. There are seven types within the species of which five affect humans and are found in soil, mud, animal faeces and animal vectors.
3. *Salmonella*: over 2000 different serotypes of *Salmonella* can infect humans, causing fever, diarrhoea and vomiting. The source of the organisms is predominantly associated with animals either directly or via contaminated sewage or wastes. The organisms are not heat-resistant and their presence in aseptically processed food is indicative of post-processing contamination. Normally a fairly large number of organisms is required to constitute an infective dose, thus contamination is invariably followed by a period of proliferation within the product before food poisoning will occur. There is current interest in the increasing incidence of *S. enteritidis* in the UK associated with eggs and poultry.
4. *Staphylococcus aureus*: this species causes a foodborne intoxication due to the growth of large numbers of organisms that produce a poisonous substance in the food. Since the enterotoxin is fairly heat-stable, food may still be toxic in the absence of viable organisms that may have been killed during processing. The source of the organisms is normally humans as the bacteria are found on the skin, in the nose and in skin lesions such as boils.
5. *Bacillus cereus*: this is a spore-forming organism that can cause two distinct syndromes. A diarrhoeal form is due to a heat-stable toxin and is associated with sporulation of the organism. An emetic or vomiting form is due to a different toxin. The spores are frequently found in cereals and other foods. Other members of the *Bacillus* group have also been implicated in food poisoning outbreaks.
6. *Campylobacter jejuni*: this pathogen will survive refrigeration and freezing but is killed by moderate heating. Food poisoning is due to an infection, although a toxin may also be implicated. The organism has been associated with raw milk, poultry and water. Person-to-person spread is known, as is the infection of domestic cats and dogs.
7. *Yersinia enterocolitica*: enteritis is the predominant clinical form of infections by *Yersinia*. Many species of animals are important reservoirs of infection and the organism has been isolated from dairy products, chocolate, meats, including vacuum-packed products, and vegetables.
8. *Listeria monocytogenes*: listeriosis is a disease transmitted from animals to human beings causing septicaemia, meningitis or abortion. Food may be implicated but it is more correctly classed as a food-carried infection. The organism can multiply, at low temperatures, in meat pastes, acid foods such as coleslaw and soft cheeses, and other foods (Lund, 1990b).

The above examples are by no means a complete list of food-poisoning organisms. Many other bacterial species have been associated with food-poisoning outbreaks but the species described are intended to indicate the different types of problems and some of the ways pathogenic microorganisms enter the food chain. For more general descriptions of food hygiene and food poisoning, reference should be made to various standard texts (Hobbs and Roberts, 1987; Singleton and Sainsbury, 1987; Doyle, 1989). The microbiological safety of food in the UK was reviewed recently and useful reports have been produced (HMSO, 1990, 1991).

8.3 Raw materials

A high-quality processed food cannot be produced from poor-quality raw materials. Uncontrolled microbial activity in a food prior to processing can cause changes in the taste, odour, texture and colour of the food. In addition to the organoleptic changes due to the breakdown of food constituents such as fats and proteins, there is also the possibility that bacterial and fungal toxins (mycotoxins) can be formed that will be stable and pass into the finished product. Botulinum toxin is normally produced by *Clostridium botulinum* and is relatively heat-labile; it can be inactivated by heating to around 80°C for 10 min, or other time/temperature combinations. However, the enterotoxin produced by *Staphylococcus aureus* can withstand boiling for up to 30 min and *Bacillus cereus* emetic syndrome is due to a toxin that can be heated to 121°C for 90 min without being inactivated (Lund, 1990a). Aflatoxins, a group of mycotoxins produced by *Aspergillus flavus* and *A. parasiticus*, are extremely heat-stable. Several bacteria, particularly members of the *Pseudomonas* group, produce heat-stable enzymes that can retain their activity after processing even though the organism itself has been killed. Such enzymes are well known in pasteurised dairy products and can cause unacceptable flavour changes in some cheeses.

Much attention should be paid to the selection and quality monitoring of the raw materials going into aseptically processed foods. It should be remembered that materials such as cereals may not necessarily be visibly mouldy if they have been cleaned before sale but could still contain mycotoxins. Inspections should be made as far back down the food chain as possible to ensure high-quality products.

Raw materials will inevitably support a microbial population and it is important to prevent the multiplication of these organisms during storage and recipe formulation prior to processing. There are a variety of ways this can be achieved.

1. *Refrigeration.* Storing produce at less than 10°C will slow, but not eliminate, spoilage problems.
2. *Freezing.* Storage at less than 0°C will kill some organisms and others will be dormant. Many enzymes and toxins remain active during frozen storage so that foods may slowly deteriorate in quality (Hobbs and Roberts, 1987).
3. *Dehydration.* Reducing the water activity (a_W) in a material will limit

microbial growth. Reducing the water activity can be achieved by evaporation of the water, or by making the water unavailable to microorganisms by the addition of salt or sugars.

4. *Preservation*. There is a trend away from the use of artificial preservatives in food but they still have a valuable role in the prevention of food spoilage and food poisoning. The use of preservatives in foods may be subject to governmental controls, which may vary from one country to another. The decision to use a preservative should be considered carefully but it is just as important to evaluate fully a move to produce a preservative-free product, as this may have repercussions on the efficiency of the food-manufacturing process.
5. *Pickling and fermentation*. The addition of acids to food improves the keeping quality by lowering the pH and inhibiting growth. In a fermentation process other antimicrobial substances are produced in addition to acids, including alcohols, carbon dioxide, antibiotics and hydrogen peroxide.

Many of these methods of food preservation interact with each other and have synergistic effects.

8.4 Processing

Aseptic processing of particulate foods invariably involves a heat-treatment process. A discussion of the use of irradiation is not covered in this chapter. The destruction of microorganisms by heat has attracted intensive research and has been the subject of other reviews (ICMSF, 1980). The main concerns are the resistance of spores and the inactivation of *Clostridium botulinum* in its vegetative and sporing states.

The heat resistance of microorganisms is dependent upon many factors. There are inherent differences in resistance between species and even strains within species. Spores are more resistant than vegetative cells, and some spores can survive for several hours at 100°C. Most vegetative cells are killed at 70–80°C in a few minutes and all are killed at 100°C in moist foods. The number of cells present is a significant factor as the death rate is a logarithmic function so that the higher the number of cells initially present, the more survivors there will be in a given time.

Heat resistance of microorganisms is influenced by the environment they are subjected to during growth, the age of cells, the growth temperature and the medium in which the microorganisms are growing. The effect of the age of cells on heat resistance is unclear, as conflicting reports have been published. In some research, young spores are said to be more resistant than old ones and *vice versa* in other reports. In general, microorganisms are at their most vulnerable during the logarithmic phase of growth when cell division is at its greatest. Since the temperature at which the bacteria are growing affects their heat resistance, it is advisable to keep food at as low a temperature as possible prior to processing to prevent the build-up of heat-resistant variants (Brown, 1989). The availability of nutrients in the growth substrate may increase or decrease heat resistance, but it is impossible to generalise and each product would have to be evaluated.

The effects of heating will be influenced by pH, water activity of the medium, the type of food, the presence of salts and many other factors. The pH of the food is probably the most significant factor, with more heat resistance being exhibited at a higher pH of 6–8, although there are some exceptions. A reduced water activity generally increases heat resistance while nutrients such as fats have a protective effect on microorganisms and sugars have been shown to protect *Escherichia coli*.

Many of these parameters will interact and multi-factor experiments may be required to establish the relationship between microbial physiology, environmental influences and heat resistance.

The heat resistance of microorganisms is calculated and expressed in a variety of ways and these are defined as follows (Singleton and Sainsbury, 1987; Gill, 1990).

1. *Commercial sterility*. Commercial sterility of thermally processed food is the condition achieved by the application of heat, sufficient alone, or in combination with other appropriate treatments to render the food free from microorganisms that are capable of growing in the food at the normal non-refrigerated conditions that the food is likely to be held during distribution (Codex Alimentarius Commission, 1979). It follows from this that the food may still contain some microorganisms but these would be inert under normal conditions. The food is not necessarily sterile in the strict sense.
2. *Pasteurisation*. This term is often somewhat loosely used. It originally indicated heating conditions that ensured the destruction of pathogenic, vegetative organisms, normally at temperatures of less than 100°C—particularly in the dairy industry. For most aseptically produced foods, higher temperatures are required to control spoilage organisms and spore-forming bacteria. In some cases the term is used to indicate a process to achieve 'commercial sterility'.
3. *Decimal reduction time* (D_{10}) *value*. This is the time required, at a given temperature, to reduce the number of viable cells or spores by 90% of the initial number. Sometimes the temperature at which the D value has been determined is shown as a subscript, in other words D_{112} means that the tests were at 112°C.
4. Z *value*. The Z value is the increase in temperature, in degrees Celsius, required to achieve a ten-fold decrease in the D value.
5. F_0 *value*. The F_0 value is the time, in minutes, required at 121.1°C to inactivate a population of cells or spores, calculated from the time required for inactivation at another temperature and assuming a Z value of 10°C. Other ways of expressing F_0 values are used depending on the Z values.

6. *12D process*. Heating a food at a given temperature for a period equal to twelve times the *D* value at that temperature.
7. *Botulinum cook.* A term frequently used to indicate a 12*D* process in respect of the endospores of *Clostridium botulinum*. The D_{121} value of *C. botulinum* spores is generally taken to be approximately 0.21 min.

The aseptic production of food involves a specified thermal process, which is controlled and monitored. Subsequently the food is filled into containers that have been suitably sterilised using filling equipment that is effectively cleaned and sterilised and maintained in a sterile condition. The containers are sealed to prevent contamination.

8.5 Aseptic packaging

Aseptic processing is aimed at producing sterile food that is placed into sterile containers under aseptic conditions. While many packaging materials will be sterile immediately after production, by the nature of the manufacturing process, they are easily recontaminated by dusts and handling during storage and prior to use. As already described, it may not be possible to achieve sterility but several processes have been developed to commercially sterilise or decontaminate the packaging materials immediately prior to the filling process. These range from tin cans that are sterilised by steam, then filled and closed in a sterile chamber, to flexible containers that are decontaminated using methods such as ultraviolet (UV) light, alcohol or hydrogen peroxide. Materials like polythene or polystyrene containers are extruded at temperatures in excess of 200°C and are sterilised by the heat of the extrusion and forming process. The containers are then filled immediately in a sterile area (Gill, 1990). The following section considers methods for the decontamination of packaging materials.

8.5.1 Hydrogen peroxide and peracids

The use of hydrogen peroxide and peracetic acid as disinfectants or sterilants has been known for many years and the bactericidal, sporicidal and fungicidal properties have been described (Baldry, 1983). The antimicrobial action of hydrogen peroxide is thought to be due to its reduction to the highly reactive hydroxyl radical which can react with membrane lipids and nucleic acids (Singleton and Sainsbury, 1987). The effectiveness of hydrogen peroxide is, as expected, influenced by temperature, with higher activity being found at higher temperatures. It has been used as a disinfectant at about 0.2% but can be used as a sterilant at 15–30% on inanimate objects (Gill, 1990). Spore inactivation of a heat-resistant *Bacillus* strain, *Bacillus subtilis* SA22 spores, in a few seconds requires a temperature in excess of 70°C at a concentration of 30% or more of hydrogen peroxide (Cerny, 1989). It is likely that the type of spores usually found on the surfaces of packaging materials are inactivated more easily than this resistant strain; however, these parameters provide a suitable safety margin for operational use.

Hydrogen peroxide is active in both the liquid (generally in aqueous solution) or vapour phase; this helps to overcome some of the problems of ensuring adequate coverage of the surface of a packaging material—particularly if the material is hydrophobic or the hydrogen peroxide is applied in an aerosol form. Hot-air drying of the packaging after hydrogen peroxide treatment has been observed to ensure sterility of the total surface (Cerny, 1989). In addition, the efficiency of sterilisation by hydrogen peroxide has been shown to be enhanced by ultrasonic energy at 16 kHz or higher (Ahmed and Russel, 1975) or in the presence of copper ions (Singleton, 1987). The combination of hydrogen peroxide treatment with ultraviolet irradiation is particularly effective. A system of generating hydrogen peroxide 'fogs' using ultrasonic generators (so-called geysirs) has been investigated, whereby the very small droplets created are attracted to the packaging surface by an electrical charge (Cerny, 1989).

In all forms of disinfection/sterilisation, microorganisms have been shown to be protected to a greater or lesser degree by the presence of materials such as fats and proteins. This protection may be due to some physical barrier or to a chemical interaction. It is essential, therefore, to consider the nature of the packaging material and also to ensure that it is maintained in a clean state and not contaminated by extraneous chemicals, dust or residues, such as sweat left on the material during handling.

The peracids (acids containing the peroxy group, –O–O– have been used as sterilants and can be regarded as derivatives of hydrogen peroxide. The most effective is peracetic acid ($CH_3.COO.OH$), which has been shown to be a bactericide, fungicide and sporicide at concentrations of 1% or less—especially under acidic conditions. It has been used in aqueous solutions and as an aerosol or vapour. Peracetic acid has a pungent smell as a 40% solution and is corrosive towards many metals. It is toxic, causing skin damage or eye and respiratory irritation and has a rat LD_{50} (lethal dose: the dose of the chemical that would kill 50% of a rat population under controlled conditions) of 1540 mg/kg (Baldry, 1983). However, its rapid action at low concentrations has made it attractive for certain applications. It has an activity similar to sodium hypochlorite, except that there are fewer residual problems since the breakdown products, acetic acid, hydrogen peroxide, water and oxygen, are all volatile (Hugo, 1971). The possibility of acetic acid (vinegar) type taints in some products should be considered.

8.5.2 Ultraviolet irradiation

Ultraviolet (UV) light has wavelengths between 328 and 210 nm but has the maximum antimicrobial effect between 280 and 250 nm, at which wavelengths nucleic acids have their maximum absorbency.

The use of high-intensity UV lamps has been known for many years and has found various applications such as for water and air sterilisation or disinfection.

UV radiation has a wide spectrum of activity and most microorganisms are inactivated within a few seconds, although, in general, Gram-positive bacteria require five times the dose of Gram-negative organisms, while spores may require ten times more. Mould spores are more resistant and require up to 50 times the dose needed for Gram-negative bacteria. This is thought to be due to the protective influence of the colour of the cell walls, which may act as a UV filter.

The antimicrobial action is due to the absorption of UV by the bases in DNA, which can induce the formation of a variety of photoproducts leading to mutation or cell death. The effectiveness of UV irradiation is not improved to any great extent by increasing temperature between 5°C and 37°C, although increasing air humidity influences the inactivation of bacterial spores.

UV irradiation has been used in aseptic packaging processes both for air sterilisation and for packaging materials, sometimes in combination with hydrogen peroxide; however, it has several disadvantages. UV light has poor powers of penetration so that it is readily absorbed by solids and can only penetrate liquids to a limited extent. Microorganisms can be easily protected by dust particles or within microcolonies. The shading effects of the geometry of the packaging material have to be considered. Of considerable significance is the ability of microorganisms to survive UV treatment. There are so-called 'dark survivor' organisms, which survive the irradiation in the first place and also 'light survivors', which are able to repair DNA damage, caused by previous exposure to UV, in visible light (300–600 nm). This photoreactivation may take a little time, so the failure to recover microorganisms immediately after treatment does not necessarily indicate sterility.

8.5.3 *Ionising irradiation*

Ionising radiations, such as gamma rays, have been used for the sterilisation of prepacked medical, pharmaceutical and biological materials. Presterilisation of packing materials in purpose-built units has some application, provided that the materials are subsequently handled and used under aseptic conditions. Ionising radiation may affect the materials being sterilised and some plastics may be damaged (see Chapter 3 on Packaging Materials). Ionising radiation from high-energy electron beams can be used on-line, since the amount of screening required is much less, and can be used on flat packaging materials where the geometry of the pack does not interfere with the application of a uniform dose of radiation (Gill, 1990).

8.5.4 *Thermal treatment*

While it is feasible to sterilise packaging materials by moist or dry heat, there are several problems to be overcome. These include damage of the packaging material, time for temperature equilibration of the packaging, condensation of water on the packaging material in the case of moist heat and the thermal tolerance of various microorganisms. In general, dry heat requires very high temperatures; for example, glass bottles have been sterilised at 270°C within 12 s but there are considerable problems in preventing recontamination during cooling and in stopping the breakage of the glass.

Moist heat at 140–160°C has more application—particularly for high-acid foods where a process akin to pasteurisation is sufficient (Cerny, 1989). High-pressure steam has been used to create a sterile environment around the filling head during filling and sealing to ensure an aseptic environment.

8.5.5 *Integrity of packs*

Even when the processing conditions are such that the product is commercially sterile and is packed into previously sterile material, the system will be severely compromised if the integrity of the pack cannot be maintained. Even minor damage to packs will be sufficient to lead to problems. Many species of bacteria have a diameter of approximately 1–2 μm and many are motile and are able to move in water-films by means of flagella or gliding motility. In addition, as products cool, the contents will contract and external contaminants could be drawn through pinhole damage to packs and recontaminate the product. The integrity of the pack can be broken in several ways. Packs become faulted during production, for example aluminium foil is particularly prone to pinhole formation at manufacture and during handling, and is therefore normally used as a laminate with paper board and low-density polythene. The sealing of the pack may be incomplete because of poor setting of the machinery or the wrong choice of materials. Pack seals can be bridged by a liquid film and this is particularly evident when handling certain foods. Sealing through the product to avoid the presence of a headspace is unlikely to be effective with food containing solid matter because the seal will be incomplete (Gill, 1990). Misaligned or dripping filling heads can cause problems of smear over the surface to be sealed, leading to subsequent microbiological contamination.

Microbiological invasion of products occurs all too frequently due to damage of the packs after filling. Poor handling of packs on conveyors and during cartoning or by stacking products wrongly, may cause the rupture of pack seals under pressure. The jostling of caps during transit has been known to cause them to loosen. Temperature fluctuations can weaken or loosen seals and this may be of particular significance if products are moved from temperate to tropical parts of the world, in which case different packaging technologies may be required. Problems with loss of pack integrity will obviously be of major significance if aseptic processing and packaging has been combined with a gas-flushed or modified atmosphere approach to extending shelf-life.

8.6 Validation of aseptic processing and packaging

In the USA, several different regulations exist for the control of plants producing aseptically processed foods; they depend upon whether the product contains meat, is an acidified food (i.e. with an equilibrium pH of 4.6 or less) or a dairy product. The common thread in these regulations is that the food processor must file with, and receive acceptance from, the appropriate regulatory authority before commencing processing (Bernard *et al.*, 1990). Similar requirements do not exist in the UK at the present time, but it is clearly essential to completely validate the efficiency of a particular process before commencing commercial production. The equipment must be capable of delivering a commercially sterile product and be adequately cleaned after production. Also, the plant must operate under aseptic conditions.

8.6.1 Plant commissioning

It is essential to ensure that microbiological considerations are taken into account from the initial design stage and it is advisable for a specialist to be part of the design team.

The equipment must be designed with effective barriers that will prevent microbiological contamination. These include steam seals around pump shafts and valve stems, or by means of positive internal pressure provided by a continuous flow of sterile air, or other gas, through the aseptic zone. Sterile air can be produced by filtration through high-efficiency particulate air (HEPA) filters. In some cases, the whole aseptic zone is filled with superheated steam. Having created an aseptic area, particularly for product packaging, it is essential that this is not breached by careless operator actions, for example when adjusting the equipment. If regenerative heating and cooling of the product is used, the pressure of the product must always be higher than the raw materials, or coolant, to prevent cross-contamination in the event of component failure.

The small size of microorganisms means that a rough surface will provide areas for attachment and colonisation. Stainless steel should be of a high grade and finished to a smooth standard. Several American sanitary standards including the 3-A and E-3-A series, which give additional guidance on this topic, have been published in the *Journal of Food Protection* since 1971. Welds are particularly prone to harbouring contaminants and these should be as smooth as possible. Connecting flexible piping to stainless steel creates an ideal environment for microorganisms—especially if the plant operates under fluctuating pressures so that gaps under joints open and close to allow ingress of nutrients and bacteria. Dead-ends in pipework, sampling points that are not flush with the surface, pistons behind valves and around seals all create areas where debris can accumulate and be difficult to clean. Microbiological colonisation will follow. The hardness of the water used for cleaning should be considered in relation to scale build-up in the plant, which creates conditions suitable for microbial colonisation.

Assuming that the plant has been designed on sound hygienic principles, it is necessary to ensure that the plant and equipment are installed to design specifications and changes are not made that will adversely affect the sterilising efficiency of the process. All items such as gauges, valves and thermometers should be checked for accuracy, and items of equipment run under actual operating conditions

Table 8.1 Test organisms most commonly used in challenge testing

Sterilisation medium	Organism
Superheated steam	*Bacillus stearothermophilus* (1518)
Dry heat	*B. polymyxa* (PSO)
H_2O_2 plus heat	*B. stearothermophilus* (1518), *B. subtilis A* or *B* or var. *globigii*
H_2O_2 plus UV radiation	*B. subtilis A*
Gamma irradiation	*B. pumilus*
Dry heat	*Clostridium sporogenes* (PA 3679)

and in 'worst-case' situations. Details of plant and equipment can be found in other reviews (CFDRA, 1989; Bernard *et al.*, 1990). The operations should be specified in a hazard analysis: critical control point system (see section 8.8.1).

8.6.2 Sterility testing

Once the plant is apparently functioning correctly it is necessary to validate the process by ensuring the product is of the desired microbiological quality. A variety of methods are used to ensure that the process can achieve commercial sterility.

8.6.2.1 Challenge testing. Challenge testing involves seeding the system with microorganisms and checking to ensure that they are successfully eliminated during the normal operation of the plant. It is clearly inadvisable to use a known pathogenic organism so the test microbe is selected to mimic potentially harmful or spoilage organisms. When selecting a microbial challenge organism, consideration is given to: (i) the types of sterilant used and the resistance of the microorganisms; (ii) the acidity of the medium; (iii) the heat resistance of the organism; (iv) the potential spoilage organisms including yeasts, moulds and bacteria; and (v) the rheology of the product in relation to the residence time in different parts of the plant. A selection of test organisms most commonly used is listed in Table 8.1 (Bernard *et al.*, 1990; Gill, 1990).

The test microorganisms can be introduced into the process in a variety of ways including drying onto surfaces especially packaging materials, attaching impregnated discs or foils within the plant and by incorporation into the product. A system of value for particulate foods is an alginate bead technique.

8.6.2.2 Alginate particle technique. Assessing the microbial lethality of a process depends upon an accurate knowledge of the temperature distribution throughout the product. If the product is a homogeneous liquid it is relatively easy to model and measure the temperature gradient and to ensure that all the of the product reaches the required temperature. This is not so straightforward with particulate foods.

Under static conditions the temperature attained at the centre of a particle can be measured using thermocouples but this is impossible under continuous-flow conditions. Alternatively, biological indicators, such as known strains of bacteria and fungi, can be used to demonstrate the severity of the process in relation to the heating temperature and heat resistance of the microorganisms.

An alginate particle technique has been developed to mimic conditions within food particles so that microbial challenge testing can be used as a predictive model for the behaviour of microbial pathogens and spoilage organisms under commercial conditions (Gaze, 1989). In this method, the indicator microorganism is mixed with the food product under test and immobilised with alginate. This mixture can be reformed into various shapes and sizes to simulate the normal particles of the product. This ensures that any reduction in numbers of the microorganism under test is due to the destruction of the microbes and not loss into the liquid matrix. It also ensures that the microorganisms are maintained in a similar environment in relation to such factors as pH, water activity and protection by food constituents such as proteins and fats, as would be found in the food product.

The technique has been utilised for predictive modelling for research purposes but also in industrial applications, such as ohmic heating, to evaluate new products and processing equipment. A variety of test organisms have been studied including:

1. *Zygosaccharomyces bailii*: yeast—tested at 60–65°C.
2. *Clostridium butyricum*: bacterial spores—tested at 80–100°C.
3. *Bacillus stearothermophilus*: thermotolerant bacteria—tested at 120–140°C

Various food products have been tested such as meat, potato, pea, carrot and peach containing particles. This technique has considerable potential for process verification and can be adapted to demonstrate a range of lethal effects under varying conditions.

8.6.2.3 Product testing. In the initial stages of plant operation, intensive sampling programmes of normal products are warranted (see section 8.7.1.1) but it is also possible to increase the confidence in the product by using modifications to the normal product or method of plant use.

Instead of the normal product, a microbiological growth medium can be used such as milk containing 0.2% glucose and an indicator such as bromocresol purple, which will change colour with shifts in pH due to microbiological activity (CFDRA, 1989). This medium is processed and packed in the usual way and the packs incubated before being checked for microbiological contamination. An alternative to packing the medium is that it remains *in situ* within the plant and the plant itself acts as the incubator. After incubation for a few days the plant is carefully broken apart and points where contamination has occurred can be identified. Such a strategy can be adopted with the product itself, bearing in mind that not all contaminants would necessarily produce visible spoilage and some materials, such as spices, in the product may have bacteriostatic effects (CFDRA, 1989).

8.7 Quality control

8.7.1 End-product analysis

The food industry has, for many years, taken samples of the finished product and checked these against a quality specification, which would include an estimate of microbiological contamination. The objective, when producing an aseptically processed food, is that the product is commercially sterile with a zero defect level; however, for many operational or economic reasons this is unlikely to be achievable and the decision has to be taken as to the level of contamination that will be commercially acceptable. This level of contamination should be such that it would be highly improbable (virtually nil) that the product contains organisms of public health significance, and the number of packs that show spoilage, or organoleptic changes, would be low.

A quality assurance strategy based upon end-product analysis alone has several weaknesses, which include: (i) the problem of taking representative samples; (ii) the detection of low levels of contamination in the product; (iii) the time that the analysis will take in relation to the product manufacturing throughput; and (iv), in the event of a faulty product being detected, the difficulty of associating the contamination with a particular component failure in the process.

8.7.1.1 Sampling. A zero defect level is desirable but probably unattainable. The commercial decision has to be taken as to the level of contamination that can be tolerated without the business suffering adverse comment and loss of sales. It is likely that the acceptable level of contamination will be 1 in 1000 packs or lower and the statistical probability of detecting low levels of contamination has to be considered. In order to detect one faulty product in 1000, with a 95% probability of success (i.e. the faulty product would be found 19 times out of 20) would require the examination of approximately 3000 samples. Even then there is still a one in 20 chance that a faulty batch would escape detection and be released. It is highly unlikely that this level of sampling and analysis can be undertaken routinely. The loss of product as a result of destructive testing would be appreciable and the cost of laboratory analysis would be prohibitive. The problem is exacerbated if it is deemed that lower levels of contamination must be detected so that a one faulty pack in 10 000 would require the examination of about 30 000 samples and so on (ICMSF, 1986).

Having taken the requisite number of samples it is also necessary to consider the distribution of contamination within the sample. As few as one or two organisms per pack are sufficient to initiate spoilage or health problems to the consumer if the conditions are such that the microorganisms can grow within the product prior to consumption.

Growth within the intact product is one thing; customer abuse is another. In the case of an outbreak of *Salmonella* food poisoning from contaminated dried milk formula in 1985, babies were being infected to such an extent that the UK company concerned attracted so much adverse publicity that it was forced into liquidation. An intensive sampling and analysis programme was conducted and it was concluded that the level of contamination of *Salmonella* organisms in the faulty packs was as low as, on average, 1.5 organisms per 450 g pack (Gilbert, 1987). This illustrates two main points. Firstly the method of subsampling of intact packs in order to ensure detection of such low levels of contamination has to be considered. In general, this is best approached by some form of pretreatment of the sample to increase the number of organisms to a level that can be easily detectable by the laboratory analysis. Secondly, it is probable that the way in which the milk formula was used would have contributed to the problem. A parent making up the formula, and leaving it warm for some time in readiness for a night-time feed, would have unwittingly created conditions suitable for the multiplication of the organism until the numbers reached an infective dose level. The food processor cannot control the actions of the consumer although the food industry has an obligation to give clear, unambiguous instructions for product use and to play a part in the general education of the consumer in food safety matters.

8.7.1.2 Time and expense. Techniques for the detection of microorganisms are time consuming. Microbiological methods are discussed later but, in outline, a count of mesophilic microorganisms in a sample using a traditional growth medium could take about 3 days. A cold enrichment technique for *Listeria* bacteria could take several weeks. If low levels of contaminants are expected, a pretreatment stage must be used to allow the organisms to multiply to a detectable level. The pretreatment may be the incubation of the entire product at an elevated temperature to encourage growth in the product, or by placing a subsample into some form of growth medium. In either case the pretreatment will take several hours or days. Even so-called rapid methods take time and can rarely be used for on-line control of the process. While an ATP determination (see section 8.7.2.4) may only take a few minutes to complete, this assumes that the number of microorganisms present is sufficiently large to come within the sensitivity of the technique. As the level of contamination in aseptically processed foods is likely to be low, a pre-incubation stage will be required.

With some products, it is possible to have a positive-hold system whereby the product is not cleared for release until the analyses are completed. This is not always feasible with short shelf-life products and demands additional storage space in all cases. The investment in staff time, laboratory premises, equipment and consumable items makes microbiological analysis relatively expensive and so it is imperative that the best use is made of laboratory facilities; this is discussed in more detail in section 8.7.2.

8.7.1.3 Corrective action. In addition to the statistical problems of taking and testing representative samples, errors are inherent in any analytical technique. False positive and negative results will occur. A false positive will impose an unnecessary financial burden on the company, while a false negative will create an unjustified sense of security. Good laboratory practices, well-trained staff and the correct choice of techniques can minimise, but can never eliminate, laboratory errors.

Once a faulty batch of product has been identified it may be possible to quarantine or recall that suspect batch and it can either be destroyed or reworked as appropriate. However, a positive isolation of a microorganism of concern rarely gives any indication of how or why the contamination occurred. It could have been the result of inadequate processing, contaminated packaging, poor cleansing of the equipment, malicious tampering by a disgruntled employee or a multitude of other reasons. It is possible to carry out some tests to help to pinpoint the fault; for example, the coliform group of organisms are unable to withstand the high temperatures normally encountered during processing and so the presence of these organisms in a product is indicative of contamination after heat treatment. Enzyme assays may also be used. The degradation of the phosphatase enzyme in pasteurised milk is used to indicate that the milk has been adequately heat-treated and that the correct pasteurisation temperature has been reached. Enzyme assays in other products may give indications of processing conditions and be of some value. A positive sample is usually a prelude to further intensive action in order to pinpoint the nature of the fault in the process.

It should also be noted that microbiological analysis by its nature invariably involves the growth of microorganisms to aid detection. This in itself poses an additional risk to other products, however small, if potentially pathogenic or spoilage of product cannot be adequately contained within the laboratory. For this reason, it is good practice for the laboratory personnel and facilities to be isolated from the main manufacturing site.

8.7.2 *The role of microbiological analysis*

8.7.2.1 Sampling. Despite the limitations described in the preceding sections, microbiological analysis is an essential tool in order to ensure high-quality, safe products. The effectiveness and value of the techniques can be improved by instituting a considered monitoring regime. Microbiological analysis has a function in the verification of good manufacturing practices but should not be used for quality control purposes in isolation. Laboratory analysis should be combined with effective on-line controls.

A full, statistically valid, sampling and testing programme may be warranted in the initial set-up of the process, to ensure compliance with the design specifications and commissioning of the plant and equipment. It may also be required when a problem has occurred, such as a food-poisoning incident, which has to be fully investigated. It is probable that only a limited number of samples will be required for the detection of a major process failure.

For routine use such an approach is inappropriate and a non-random sampling plan should be adopted. When designing such a plan consideration should be given to:

1. Taking samples at different stages in the process;
2. Taking samples from food-contact surfaces by means of swabs or rinses to confirm the effectiveness of the plant-cleaning programme;

3. Looking for the 'worst-case' situation such as the first product through the plant on each production run, or the build-up of contaminants in the plant over time or a known weak spot in the process, and sampling at these times or places;
4. Taking samples to coincide with particular events such as line stoppages;
5. Pre-incubating finished products or bulk samples at elevated temperatures and checking for pack deformation such as blowing, shifts in pH or redox potential or organoleptic changes, to guide sample selection. While this is a valuable technique, consideration has to be given to the type of product, preservatives used, temperature of incubation and time to ensure conditions conducive to the growth and detection of the type of contaminants that are anticipated.

Sampling from operational plant is best limited to points of ready access to the product such as in balance tanks. Taking a sample should be regarded as a potential contamination incident since the process has been interfered with in some manner. Samples should only be taken by fully trained operatives using aseptic techniques and sterile equipment. Various on-line sampling techniques have been developed using diaphragm and syringe systems or aseptically operating valves or sample deposition into sections of the plant that can be removed for analysis. However, all these methods are prone to contamination and their use should be fully evaluated and controlled before they are used routinely.

8.7.2.2 Sample handling. It is essential that contaminants are not introduced into packs during opening. Dirty packs should be cleaned with a suitable detergent and dried. The area to be opened can then be decontaminated with a volatile disinfectant such as acid alcohol or 2% peracetic acid (CFDRA, 1989), which will not leave residues on the pack which may be introduced into the sample. Disinfectants having bactericidal rather than bacteriostatic effects must be used, so low concentrations of quaternary ammonium compounds (QAC) or phenolic materials should be avoided. If taste assessments are to be made in addition to microbiological analysis, the sample should be opened in an area free from microbiological cultures, and pungent materials must be excluded from the room. In particular, phenolic compounds and sodium hypochlorite solutions should not be used, as these react together to form chlorophenol-type compounds, which are readily absorbed into foods and can cause taint at levels of about one part per billion. Other absorbed taints have been described (Whitfield, 1983; Goldenberg et al., 1975).

Packs should be opened by cutting or piercing the surface with a sterile instrument. It is advisable to enter through the base of the pack so that seals are left intact in order that they can be checked for faults in the event of sample failure, and also to minimise the risk of contaminating the product. Samples should be opened in a clean environment that has been previously disinfected and the area fed with clean, filtered air. Particular care must be taken when handling samples showing evidence of pressure build-up and swelling. After opening, a test sample and a reference sample should be withdrawn. The reference sample should be stored at less than 7°C, but not frozen, so that a repeat analysis can be made if required. It may be necessary to store a subsample at 25°C if thermophilic organisms are suspected.

With particulate foods, it is unlikely that the contaminants will be evenly distributed throughout the pack and it is essential that a large sample is taken and thoroughly stomached (a way of separating microorganisms from samples), or homogenised, to ensure a representative sample. Alternatively, known problem components such as large particles, or the material adjacent to the seal should be selected for analysis. All subsequent laboratory operations should be undertaken with due regard to aseptic techniques and using recognised methods, details of which can be found in a range of microbiological texts (ICMSF, 1978). The following paragraphs highlight particular points of relevance to aseptically processed particulate foods.

8.7.2.3 Microbiological methods. Routine analysis of samples should concentrate upon simple, reliable techniques. There is little point in looking for specific pathogens, which often require complex methods, if the probability of them being present is low. A succession of negative results means that either the pathogen was not present or, just as likely, that the sampling was inadequate to detect the presence of low numbers of contaminants or that the testing methodology was defective. Routine microbiological techniques should concentrate upon testing larger numbers of samples by simple methods that serve to indicate potential problems and should be supported by other tests such as changes in pH or redox potential that may indicate microbial activity. Specific, complex techniques should be reserved for known problems such as blowing packs or the investigation of food-poisoning or spoilage outbreaks.

In general, for most aseptically processed foods, any contamination with microorganisms is significant as the objective is to produce 'sterile' food. Techniques should be designed, therefore, to detect rather than to enumerate microorganisms. If a finished product is known to contain organisms that are inert, under normal storage and use conditions, then the microbiological techniques must be designed to exclude those types, although in some cases the monitoring of these organisms can serve as a useful indicator of some changes occurring in the production process.

Various physiological groupings of microorganisms can be defined and techniques designed to detect groups of significance in a specific product. Aerobic organisms or facultative aerobes are grown in media under aerobic conditions, while anaerobic and facultative anaerobes can be detected in media with a low redox potential and grown under anaerobic conditions. Some strict anaerobes will be killed by an exposure to air but, in general, foodborne anaerobic species of significance are aerotolerant to some degree, and will survive transient exposure to air during sampling and testing. These groups of microorganisms can be detected by incubating about 10 g of sample in a broth culture, or if the contamination has been developed to a high enough level, the organism can be shown by streaking a small amount onto a suitable agar medium or by a modified Miles and Misra droplet technique. The temperature and time of incubation depend upon the organism being sought. Thermophilic (heat-loving) organisms will have an optimum growth temperature of greater than 45°C as distinct from mesophilic types, which will have an optimum of between 20°C and 45°C, while psychrophiles will grow best at or below 15°C and have an upper limit of growth of 20°C. The term psychrotroph is also used for an organism that will grow at low temperatures (e.g. 0–5°C) but which has a higher optimum temperature for growth between 15 and 20°C. Thermoduric or thermotolerant organisms can survive temperature/time combinations that would normally be lethal for many, or most, vegetative microorganisms. Thermoduric organisms will survive elevated temperatures but will not necessarily be thermophilic (i.e. capable of growing at high temperatures) and are likely to be mesophilic. Thermoduric psychrotrophic organisms have been described, for example in dairy microbiology, that will survive the pasteurisation process but will subsequently grow and spoil pasteurised milk stored at refrigerated temperature. A typical incubation temperature for thermophilic organisms could be 55°C for 5 days, while for mesophiles temperatures of 37°C and 30°C are recommended. Incubation may be continued for up to 14 days to allow for the repair and growth of heat-damaged organisms.

8.7.2.4 Rapid methods. Normal cultural methods rely on the growth of the microorganisms to a point where they can be visibly detected by either the growth of the colony on an agar plate or turbidity in a broth medium. The detection of growth may be reinforced by changes in pH indicators as a result of microbial activity.

Several methods, however, rely on some other parameter to indicate microbial activity. Dye-reduction tests, such as the resazurin test for dairy products rely on colour changes of a redox indicator from the blue, fully oxidised state, through pink to a colourless state as the milk substrate is reduced by microbial activity. Other methods are amenable to automation, one such being an electrical impedance method. This technique depends upon changes in electrical conductivity of a suitable medium as a result of microbial growth. If the incubation temperature is very accurately controlled, very small changes in impedance can be detected allowing more rapid detection in a fully automated system. The technique still has some disadvantages, one of the main ones being that only very small samples are used, increasing the possibility of missing contaminating organisms. Also, the bacteria have to multiply to a sufficiently high number before detection of impedance changes is possible. Since the number of contaminants in aseptically produced food will be low, the incubation period for the sample in the equipment may have to be prolonged, or the samples will have to be pre-incubated before testing. Nevertheless, the technique has considerable potential for the routine monitoring of larger numbers of samples than is usually convenient with cultural methods.

Another type of rapid method relies on the detection of adenosine triphosphate

(ATP), which is present in all living cells. The ATP is detected by triggering a luciferin-luciferase light-emitting reaction and the amount of light emitted can be related to the number of organisms present. It is sometimes difficult to differentiate between ATP in plant and animal cells and microbial ATP, and the technique will not detect low levels of contamination without pre-incubation of the sample.

Other techniques use microscopic examination, and the direct epifluorescent filter technique may have application. In this technique, organisms are extracted from the food substrate, collected by filtration and stained with a fluorescent dye, which is specific for living cells. The bacteria can then be counted using fluorescent microscopy (Pettipher, 1983). The method is rapid and relatively cheap once the equipment has been purchased but there may be problems with food debris on the slide and the sensitivity of detection.

It is likely that techniques such as these, and others in development, will have application in the microbiological monitoring of particulate foods but, before embarking on automated techniques, the sensitivity of the methods should be evaluated in relation to the time for detection, the cost of the equipment and the range of microorganisms that can be detected.

8.7.2.5 Swabs and rinses. The cleanliness of food-contact surfaces, such as in the processing equipment and packaging materials, may need to be assessed. Microbial contaminants can be harvested by the use of sterile swabs drawn over the surface but this is only practicable when the surface is relatively heavily soiled. Microorganisms can be washed off the surfaces using a sterile solution such as a buffered saline solution. The harvested organisms are then detected by normal microbiological techniques. If large-volume rinses are used, which are drawn through the whole plant, it will be necessary to concentrate the microbial contaminants by filtration through a membrane filter with a mean pore diameter of about 0.45 μm prior to testing.

8.8 Quality management

A quality management scheme would have the objective of designing and implementing a programme that will result in a high-quality product that will delight customers, minimise wastage and improve productivity. This would lead to a greater competitive edge, increased profitability and improved business prospects.

The development of a quality management programme falls into several parts, which tend to overlap. The first stage is to review and analyse the existing situation. This is approached by a systematic audit of working practices in the factory, gathering and analysing data as appropriate to define the key areas; it would involve on-site observations and discussions with appropriate management staff. This is followed by an education and communication programme so that staff at all levels share the quality objective to ensure the maximum support for any subsequent stages. Without a total commitment from the most senior manager, the programme will not be effective, as quality objectives will be too easily subverted by other production pressures, to the long-term detriment of the company.

Implementation involves the adoption of good manufacturing practices throughout the unit and, if it does not already exist, a good manufacturing practices (GMP) manual of procedures would be developed and incorporate such techniques as an hazard analysis: critical control point (HACCP) approach to product safety and quality. HACCP systems are discussed in more detail in section 8.8.1. The objective is to define procedures, allocate responsibilities and institute checks while maintaining a high level of system 'ownership' by all personnel involved. The manual would also consider supplier quality assurance procedures including specifications, monitoring and control parameters. It is also necessary to incorporate procedures for product identification and traceability in the event of problems occurring.

On completion of this phase it is probable that the system in place at the premises would satisfy many of the requirements for accreditation under ISO 9000 (ISO, 1987). Accreditation under ISO 9000 will assist in the demonstration of a 'due diligence' approach, which is a defence against prosecution under the UK Food Safety Act 1990. Development of a quality assurance programme is not a once only change but is a process of continuous evolution while monitoring and measuring effectiveness. It is necessary to plan an on-going programme including periodic independent reviews of operations and amendments to the GMP manual as appropriate.

In addition to the quality management scheme, it would be advisable to develop a crisis management capability so that even the unforeseen event does not catch the company unaware, with the disastrous consequences that could follow. A package directed at crisis preparation would include logistics planning, recall procedures and media training, backed by the quality assurance programme. Simulation exercises should be conducted periodically to ensure the programme is working effectively. Many companies in the food industry have developed quality management systems, including accreditation under ISO 9000, and have invariably reported cost savings through improved efficiency that have outweighed the costs of implementation.

8.8.1 Hazard analysis: critical control point systems

The concept of building safety into food preparation was given a major impetus by the USA space programme. The systematic approach to food quality and safety has come to be known as the hazard analysis: critical control point (HACCP) system and was developed by the Pillsbury Company in the USA. It was based upon the idea of a zero-defect strategy developed by NASA for space exploration. At the time, it was uncertain how food materials would react under space-travel conditions, nor was it known what the susceptibility of the astronauts to foodborne microorganisms would be in what was essentially an unknown environment. Reliance could not be placed upon end-product analysis and so a system of building safety into every stage of the food preparation process was developed. This is equally applicable to any aseptic food processing.

An HACCP system is designed to address the following working practices:

1. To assess the hazards to a product and decide upon the critical control measures to limit the risks;
2. To define the responsibilities of the work force in relation to the control of hazards;
3. To improve the efficiency of the production process, reduce wastage and ensure safe wholesome products; and
4. To provide the procedures and documentation to complement an ISO 9000 approach to quality management.

HACCP systems have been developed to control microbiological risks (ICMSF, 1988) but the concept can be used to control any hazard that can have a detrimental effect on the commercial value of the product. In this context hazards will include microbiological contamination, chemical adulteration, the presence of foreign bodies, physical changes and visual or olfactory abnormalities.

In outline, HACCP systems specify:

- What are the problems
- How are they controlled
- Who is responsible
- Where is the action taken
- What is that action
- How often is it performed
- How much non-conformity can be tolerated
- Who checks that the system is operating effectively.

The HACCP approach is suitable for any size of food business and should be applied at all stages in the product cycle, from the quality of the raw materials to the packaging of the final product. Fully documented HACCP systems are recognised as important aids to product safety and have been specified in various regulations and proposals on food topics issued by the European Community. It is one of the factors considered under the UK food law (MAFF, 1990) to rationalise the official inspection process of food manufacturing and processing premises. Similar schemes are in operation in the USA (Corlett and Stier, 1991).

A systematic approach to the design of an HACCP system ensures that all the relevant factors are taken into consideration, starting with familiarisation with the production process and design parameters (Mayes and Kilsby, 1989). These are organised into a flow-diagram for the process, which is checked against a range of operating conditions and amended as necessary. The process flow-chart will have defined various operations from which the hazards to the product can be identified. The relative risk to the product from these hazards can be assessed and categorised in accordance with the perceived severity (Corlett and Stier, 1991). The points at which the hazards can be controlled are identified.

Some of these points will ensure the complete control of a hazard such as a heat-treatment process, while others will be points where the risk can be reduced but not eliminated. Most hazards will be controlled by simple on-line actions, while others will require more sophisticated control procedures and specialised equipment such as metal detectors. The programme will define who will act and what they will be required to do, as well as specifying how often the checks are made by either continuous observations or periodic actions. It will define the conditions that will trigger further action. As well as on-line controls there will invariably be the need for the other checks, such as microbiological or chemical analysis, in order to verify that the system is operating correctly. No production remains static and it is essential that the programme is reviewed and modified as necessary and that staff awareness is maintained.

8.9 Further thoughts

Ever since man ceased to be a hunter-gatherer, and the increasing population started to live in communities, a battle has been waged against microorganisms so that food can be stored in the time of feast to protect against famine. Food hygiene laws have been developed by many cultures to safeguard food, based no doubt on the best form of scientific study, that is, detailed observations over time. The laws of Moses in about 2000 BC formalised observations, between the incidence of disease and food handling practices, without necessarily understanding the scientific causes. It was not until the time of Louis Pasteur (1822–1895) that mankind started to understand the role of microorganisms in food safety. Microbiology as a science has only really developed in the twentieth century and, for the most part, has been directed at the identification and control of infectious diseases.

It is much more recently that attention has been focused on the role of microorganisms in food spoilage. Since a substantial proportion of the food produced worldwide fails to reach the consumers because of spoilage, there is still much to be learnt and more to be put into practice. In the developed world, consumers demand a more varied diet combined with the convenience of ready-prepared meals with a long shelf-life. Major advances have been made in the understanding of the genetic make-up of microorganisms and we now have the ability to genetically manipulate microorganisms to perform predetermined biochemical processes. However, the full potential of such techniques will not be achieved without a greater understanding of microbial ecology and the behaviour of microorganisms in natural environments. The complex interactions of microbes in materials such as foodstuffs need much greater study.

Aseptic processing of food goes a long way to ensuring safe, wholesome products with long shelf-lives. Although food technologists and engineers will continue to develop new production methods, microorganisms are unforgiving of errors and extremely adaptable. Failure to design equipment with a clear understanding of microbiological principles will be followed by spoilage problems. Poor manufacturing practices will be revealed by contamination incidents, leading to food-poisoning outbreaks or unpalatable products. Process design must be complemented by a detailed knowledge of the behaviour of microorganisms in the heterogeneous environment within a foodstuff where factors such as pH, water activity, salt, temperature, preservatives and many more, interact to influence the survival and growth of potentially harmful microorganisms.

There is little doubt that new technologies will emerge but, make no mistake, microorganisms will always be around to exploit every available ecological niche.

References

Ahmed, F. I. K. and Russel, C. (1975) Synergism between ultrasonic waves and hydrogen peroxide in the killing of microorganisms. *Journal of Applied Bacteriology*, **39:** 31–40.

Baldry, M. G. C. (1983) The bactericidal, fungicidal and sporicidal properties of hydrogen peroxide and peracetic acid. *Journal of Applied Bacteriology*, **54:** 417–423.

Bernard, D. T., Gavin, A., Scott, V. N. *et al.* (1990) Validation of aseptic processing and packaging. *Food Technology*, **44:** 119–122.

Brown, K. (1989) UHT thermobacteriology. In *Recent Developments in Aseptic Technology*. Symposium Report, 4–6 December 1989, published by Campden Food and Drink Research Association.

Campden Food and Drink Research Association (CFDRA) (1987) Guide to the establishment of hazard analysis critical control point systems (HACCP). *Technical Manual No. 19*.

CFDRA (1989) The microbiological aspects of commissioning and operating aseptic production processes. *Technical Manual No. 24*.

Cerny, G (1989) Sterilisation of packaging materials for aseptic packaging—problems and developments. In *Recent Developments in Aseptic Technology*. Symposium Report, 4–6 December 1989, published by Campden Food and Drink Research Association.

Corlett, D. A. and Stier, R. F. (1991) Risk assessment within the HACCP system. *Food Control*, **2:** 71–72.

Codex Alimentarius Commission (1979) *Recommended International Code of Practice for Low-acid and Acidified Low-acid Canned Foods*. FAO/WHO.

Doyle, M. P. (1989) *Foodborne Bacterial Pathogens*. Marcel Dekker, New York.

Gaze, J. E. (1989) Process verification by means of an alginate particle technique. In *Recent Developments in Aseptic Technology*. Symposium Report, 4–6 December 1989, published by Campden Food and Drink Research Association.

Gilbert, R. J. (1987) Foodborne infections and intoxications. In *Microbiological and Environmental Health Problems Relevant to the Food and Catering Industries*. Symposium organised by the Campden Food and Drink Research Association and the Public Health Laboratory Service, UK.

Gill, J. M. (1990) Recent developments in aseptic processing—a literature review. *Food Focus No. 10*. British Food Manufacturing Research Association.

Goldenberg, N. and Matheson, H. R. (1975) Off flavours in food—a summary of experience 1948–1974. *Chemistry and Industry*, 5 July, 8–14.

HMSO (1990) *The Microbiological Safety of Food. Part 1*. Report of the committee chaired by Sir Mark Richmond.

HMSO (1991) *The Microbiological Safety of Food. Part 2*. Report of the committee chaired by Sir Mark Richmond.

Hobbs, B. C. and Roberts, D. (1987) *Food Poisoning and Food Hygiene*. 5th ed. Edward Arnold, London.

Hugo, W. B. (1971) *Inhibition and Destruction of the Microbial Cell*. Academic Press, London.

International Commission on the Microbiological Specification in Foods (ICMSF) (1978) *Microorganisms in Foods*. Volume 1. Their significance and methods of enumeration. 2nd ed. University of Toronto Press, Toronto.

ICMSF (1980) *Microbiological Ecology of Foods*. Volume 1. Factors affecting the life and death of microorganisms. Academic Press, London.

ICMSF (1986) *Microorganisms in Foods*. Volume 2. Sampling for microbiological analysis: Principles and specific applications. 2nd ed. University of Toronto Press, Toronto.

ICMSF (1988) *Microorganisms in Foods*. Volume 4. Application of the hazard analysis critical control point (HACCP) system to ensure microbiological quality and safety. Blackwell Scientific Publications, Oxford.

International Standards Organisation (ISO) (1987). ISO 9000 series: *Quality Systems*. (Equivalent to BS 5750:1987; EN29000:1987.)

Lund, B. M. (1990a) Foodborne disease due to *Bacillus* and *Clostridium* species. *Lancet*, 982–986.

Lund, B. M. (1990b) The prevention of foodborne listeriosis. *British Food Journal*, **92:** 13–22.

Mayes, T. and Kilsby, D. C. (1989) The use of HAZOP hazard analysis to identify critical control points for the microbiological safety of food. *Food Quality and Preference*, **1:** 53–57.

Ministry of Agriculture, Fisheries and Food (MAFF) (1990) Food hygiene inspections. *Food Safety Act 1990, Code of Practice No. 9.*

Pettipher, G. L. (1983) *The Direct Epifluorescent Filter Technique for the Rapid Enumeration of Microorganisms.* Research Studies Press (John Wiley), Chichester.

Singleton, P. and Sainsbury, D. (1987) *Dictionary of Microbiology and Molecular Biology*. 2nd ed. Wiley Interscience Publications, (John Wiley), Chichester.

Whitfield, F. B. (1983) Some flavours which industry could do without. *CSIRO Food Quality Research*, **43:** 96–106.

9 Aseptic packaging of liquid foods

S. WAKABAYASHI

9.1 Introduction

In the course of Japan's rapid growth into an economic superpower in the world, its social environment and mode of living have vastly changed. Eating habits have reached a turning point, as a result of ever-increasing participation of consumers in public affairs, together with an ageing society. Palatability, contribution to health, especially freshness, being required, there is an increased demand for such packaged foods as may readily be consumed at any time of need. In the food packaging field, maintenance of freshness and keeping quality of the packaged foods are required.

As they are sterilized at a high temperature for a short time, immediately cooled down and then filled at a low temperature, aseptically packaged foods may exhibit, for example, the following merits:

1. Less change in flavor and color tone of the product;
2. Less loss of nutrients in the product;
3. Longer shelf-life of the product, making ambient temperature distribution and planned production possible;
4. Abundant packaging materials such as paper, plastic, can and so on.

Aseptic packaging is thus being widely used and this chapter describes an aseptic filling and packaging systems for liquid foods supplied by Shikoku Kakoki Co. (SKC) of Japan.

9.2 Requirements for an aseptic packaging machine

Although the term 'aseptic packaging machine' is defined in many different ways, it is extremely difficult to prove the degree of asepsis when directed at commercial systems. The percentage of defectives is said to vary from 0.1 to 0.01%. For certain operations, 3000 packs are filled with a culture medium, incubated at 35°C for 5–7 days and all the products checked for growth of microorganisms. If the products are all cleared the result is quoted as 'no detection of defective specimen' for such testing conditions at the time of shipment of the machine. Since checking of the total number of products cannot be performed after delivery to the end users, it is essential to have some assurance that if the specified conditions are operated, all the products would be aseptic. The performance required for aseptic packaging may be summarized as follows:

Table 9.1 Outline of aseptic packaging machines from various manufacturers[1]

Name of manufacturer (Country)	Type	Packaging materials	Shape	Sterilization of container	Principal products to be filled
Tetra Pak (Sweden)	Rolled paper		Brick	35% H_2O_2 immersion at 80°C	Milk, juice
Shikoku Kakoki (Japan)	Rolled paper		Brick	35% H_2O_2 immersion at 80°C	Milk, juice
International Paper (USA)	Rolled paper	Paper,	Brick	17.5% H_2O_2 immersion	Milk, juice
Ex-Cell-O (USA)	Carton	aluminum foil,	Gable top	BQG treatment, 35% H_2O_2 Spray	Milk, cream, soup
Liquipak (Switzerland)	Carton	polyethylene	Gable top	1% H_2O_2 spray, UV Irradiation	Milk, cream, juice
Jagenbeck (Germany)	Carton		Gable top	35% H_2O_2 spray	Milk, juice
Shikoku Kakoki (Japan)	Carton		Gable top	35% H_2O_2 spray	Cream, soup
Gusti (Germany)	Cup supply	Plastic	Cup	35% H_2O_2 spray at 80°C	Dessert foods
Hamba (Germany)	Cup supply	Plastic	Cup	35% H_2O_2 spray at 80°C	Dessert foods
Benhil (Germany)	Cup supply	Plastic	Cup	UV irradiation	Dessert foods
Dai-Nippon (Japan)	Cup supply	Plastic	Cup	35% H_2O_2 spray	Dessert foods
Shikoku Kakoki (Japan)	Cup supply	Plastic	Cup	UV irradiation, 35% H_2O_2 spray	Dessert foods
Benhil (Germany)	Plastic sheet	Plastic	Cup	35% H_2O_2 immersion	Coffee cream, dessert foods
Hassia (Germany)	Plastic sheet	Plastic	Cup	35% H_2O_2 immersion, UV irradiation	Yoghurt, dessert foods
CKD (Japan)	Plastic sheet	Plastic	Cup	35% H_2O_2 immersion	Dessert foods
ERCA (France)	Plastic multi-layer sheet	Plastic	Cup	Film laminate heated at the time of co-extrusion	Coffee cream, dessert foods
Dole (USA)	Metal can	Metal	Can	Dry steam at 210–220°C	Juice, paste
	Composite can	A mixture of paper, aluminum and polyethylene	Can	Heated air	Juice, paste
Selac (France)	Metal can, bottle	Metal, plastic	Can, bottle	H_2O_2 and chlorine water	Juice
Scholle (USA)	Bag-in-box	Plastic	Bag	Gamma rays	Juice
Dai-Nippon (Japan)	Bag-in-box	Plastic	Bag	EOG treatment	Cream
Toppan (Japan)	Bag-in-box	Plastic	Bag	Gamma rays	Cream

[1]Source: Nakae, T. and Yagi, N. (1987) *Technical Manual for the Aseptic Filling Machine of Foods*, pp. 127–137. Business Centre Sha, Japan.

Table 9.2 Sterilization systems[1]

Sterilization systems	Representative manufacturers	Country
Direct heating		
Steam jet (injection)	Alfa-Laval	Sweden
	APV	UK
	Cherry-Burrell	USA
	Iwai Machinery	Japan
Liquid jet (infusion)	Brell and Martel	France
	Dasi Industries	Canada
Indirect heating		
Plate system	APV	UK
	Sordi	Italy
	Alfa-Laval	Sweden
	Ahlborn	Germany
	Iwai Machinery	Japan
	Stork	Holland
Tubular system	Cherry-Burrell	USA
	Stork	Holland
Scraping system	Alfa-Laval	Sweden
	Iwai Machinery	Japan

[1]Source: As Table 9.1.

1. That sterilization of the packaging machine is simple and sterility is maintained.
2. That sterilization of containers is assured.
3. That operation and control are readily achieved.
4. That wash-down of the packaging machine is easy.
5. That the machine problems, and so on, may readily be rectified.

The main aseptic packaging systems offered worldwide are summarized in Table 9.1. Methods for sterilizing the different product types are shown in Table 9.2. For solid products, an appropriate method should be selected, such as a scraping method or batch pressurized heating, and so on, depending on the physical properties such as viscosity, quality and solids content of the product to be filled, and so on.

9.3 Performance and features of an aseptic packaging machine

For an aseptic packaging machine: (i) sterilization of packaging materials; (ii) filling and sealing; and (iii) an aseptic positive-pressure chamber are essential requirements (Figure 9.1).

9.3.1 Sterilization of packaging materials

9.3.1.1 Method of sterilization. Sterilization of containers may be performed with steam heating, chemical agents (ethylene oxide and hydrogen peroxide),

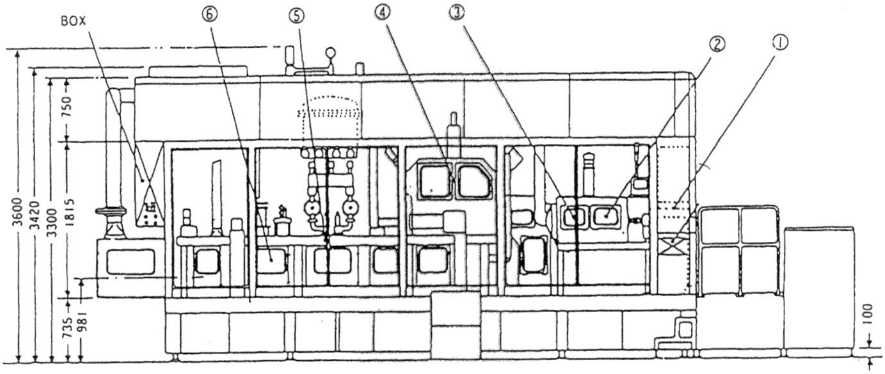

Figure 9.1 Outline of an aseptic packaging machine: 1 = carton in-feed section; 2 = sterilization section; 3 = drying section; 4 = forming section; 5 = filling section; 6 = aseptic chamber.

ultraviolet, far-infrared, microwave, gamma radiation, and so on; in most cases, however, hydrogen peroxide and ultraviolet irradiation are used.

9.3.1.2 Rolled paper. Hydrogen peroxide is generally used as a sterilizing agent, and in the case of packaging material in a form of a rolled sheet, a system (Figure 9.2) that allows it to pass through a bath of 35% hydrogen peroxide at a

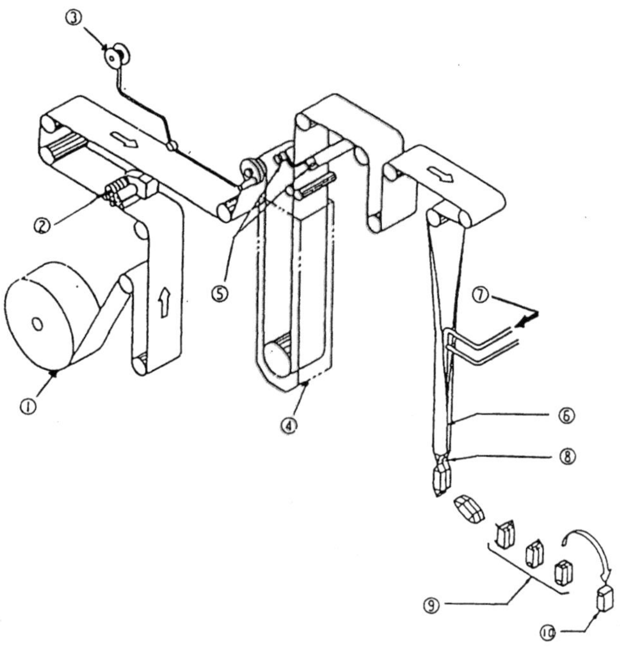

Figure 9.2 The manufacture of packs from roller paper: 1 = rolled paper; 2 = stamping; 3 = sealing tape reel; 4 = H_2O_2 bath; 5 = air knife; 6 = vertical seal; 7 = from product line; 8 = horizontal seal; 9 = folding into formation; 10 = discharge.

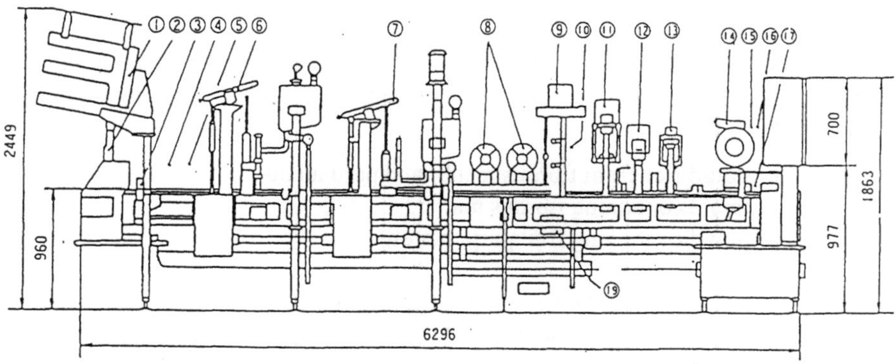

Figure 9.3 Aseptic machine for cup-type containers (ULA 120). 1 = container in-feed device; 2 = container separation section; 3 = container detection section; 4 = foreign body removal section; 5 = container sterilization section; 6 = primary filling section; 7 = secondary filling section; 8 = rewinder section; 9 = stamping section; 10 = lid material sterilization section; 11 = sealing section; 12 = pressurized cooling section; 13 = trimming section; 14 = wind reel section; 15 = skirt fold-in section; 16 = stamping section in case of jet printer; 17 = discharge section; 18 = discharge conveyor section; 19 = holder plate-changing section. Note: 4,5,7,10,15 and 16 are optional; 9 or 16 is used for stamping; 5 and 10 are extended shelf-life specifications. The machine capacity is 9000 packs/h (6 lanes × 1500 packs/h).

temperature of 75°C for more than 7 s is generally used. The peroxide is then removed by means of an air knife, and so on, using hot air.

9.3.1.3 Formed container. For a formed container such as a paper container, cup (see Figure 9.3) or bottle, hydrogen peroxide and ultraviolet irradiation are generally used as sterilizing agents. Although removal by means of drying after spraying minute droplets of hydrogen peroxide is a general practice, a combined use of ultraviolet irradiation and hydrogen peroxide is being developed, in order to enhance the sterilization efficiency. In the case of the combined use of hydrogen peroxide and ultraviolet irradiation, the intensity of the latter may be lowered because of the hydrogen peroxide film; it has become possible to secure a high sterilization level by the combined use of hydrogen peroxide of a low concentration and ultraviolet light, which produce a synergistic effect.

9.3.1.4 Unformed carton. Where the base has not been formed, two different methods can be considered, namely sterilizations prior to and after the formation of the base.

1. *Sterilization prior to formation of the carton base.* The entire carton can be sterilized in the unfolded condition to produce a more uniform sterilization. The drawback is, however, that the plant will be comparatively high in price, as the forming section after the sterilization needs to be completed aseptically.
2. *Sterilization after formation of the carton base.* This will give better control of the carton-forming step, but its demerit is that it will increase the risk of incomplete sterilization.

9.3.2 *Filling and sealing*

Products to be filled have varying viscosities, may contain fibrous matter or particles of fruit flesh and be of varying size. At present, no universal nozzle is available to satisfy all of these attributes. Three types of nozzle are employed by our company; they are:

1. For liquid free from fibrous matter and particulates;
2. For liquid containing fibrous matter;
3. For liquid containing fruit particulates.

9.3.2.1 Liquid free from fibrous matter and particulates. Meshes are designed into the tip of nozzle so that liquids may be delivered having viscosities ranging for that of water up to 200 cp. The same nozzle is used for liquids such as milk or cream. For cream, such a nozzle is popular, although care has to be taken to minimize shear to the product to prevent churning, and also temperature abuse. Compared with the other types of nozzles, its capacity and speed are greatest—particularly when the height of the containers is taken into consideration.

9.3.2.2 Liquid containing fibrous matter. Known as the 'tip-check nozzle', the tip of the nozzle is made into a trumpet shape. The filling medium can have a viscosity up to 300 cp. This type is suitable for juice, soup, and so on.

9.3.2.3 Liquid containing fruit flesh. The 'tip-spool nozzle' has the tip of the nozzle made into a spool shape. It is suitable for filling liquid up to 400 cp in viscosity. When there is a large difference in the specific gravity between fruit flesh and liquid, uneven distribution of fruit flesh will take place unless sufficient agitation is applied in the storage tank.

9.3.2.4 Other nozzles. Special designs of nozzle are provided for liquids of higher viscosities. With sensitive products, depending on the speed of liquid in the filling line, SKC will design special filling nozzles for the customer.

When mounting special nozzles onto an aseptic packaging machine, careful consideration is required to maintain CIP integrity, sterilization of the filling line, maintenance of sterility when changing product, and so on. Filling in an aseptic room is also required in order to avoid secondary contamination. CIP has to be automatic, and sterilization is usually performed by means of high-pressure steam.

9.3.3 *Aseptic positive-pressure chamber*

An aseptic positive-pressure chamber is installed between sterilization of packaging materials and filling/sealing to avoid secondary contamination. A reliable method of air filtration is necessary for this. When a high-performance filter (HEPA) is used in combination with a prefilter, particles of 0.3 μm can be separated out so that bacteria in the air may be completely removed to produce sterile air. The inside of the chamber is maintained under a positive pressure of

1 mm water to prevent any contamination from the outside. Another requirement is that all the units inside the chamber must be capable of being cleaned and sterilized. Automatic cleaning is performed by means of spray balls; hydrogen peroxide spraying and hot-air drying are generally used for sterilization. Sterility is maintained with filtered and sterile air using an HEPA filter, and positive pressure is maintained until the end of production.

Index

MIX
Papier aus verantwortungsvollen Quellen
Paper from responsible sources
FSC® C105338

If you have any concerns about our products,
you can contact us on
ProductSafety@springernature.com

In case Publisher is established outside the EU,
the EU authorized representative is:
Springer Nature Customer Service Center GmbH
Europaplatz 3, 69115 Heidelberg, Germany

Printed by Libri Plureos GmbH
in Hamburg, Germany